Conservation and Management of Neotropical Migrant Landbirds

Conservation and Management of Neotropical Migrant Landbirds in the Northern Rockies and Great Plains

by
David S. Dobkin

High Desert
Ecological Research
Institute
Bend, Oregon

Illustrated by Chris Moore

University of Idaho Press
Moscow, Idaho
1994

Copyright © 1994 by David S. Dobkin.

Printed in the United States of America

All rights reserved. No part of this book may be reproduced, stored in a retrieval system, or transmitted in any form or by any means, electronic, mechanical, photocopying, recording, or otherwise, except for use in reviews, without the prior permission of the publisher.
98 97 96 95 94   1 2 3 4 5

Library of Congress Cataloging-in-Publication Data

Dobkin, David S.
  Conservation and management of neotropical migrant landbirds in the Northern Rockies
  and Great Plains / by David S. Dobkin:
    illustrated by Chris Moore.
    p.   cm. (Northwest naturalist books)
  Includes bibliographical references (p.   ).

ISBN 0-89301-168-1

1. Birds—Rocky Mountains—Migration. 2. Birds—Great Plains—Migration. 3. Birds, Protection of—Rocky Mountains. 4. Birds, Protection of—Great Plains. 5. Bird populations—Rocky Mountains. 6. Bird populations—Great Plains. I. Title. II. Series
QL683.R63D63 1994
333.95'8—dc20   93-11355
            CIP

To Donna

for her loving insights into the human condition
that inform our view of the natural world

CONTENTS

Preface xi

Acknowledgments xiii

Introduction 1
    Identification of the Problem: Forest Fragmentation
    Tracking Long-Term Population Trends
    Problems in Tropical Wintering Areas
    The Picture in Western North America

Survey of Management Activities and Impacts in the Northern Region 9
    Coniferous Forests
    Douglas Fir and Ponderosa Pine
    Other Coniferous Forests
    Fire and Avian Community Response
    Effects of Livestock Grazing
    Impacts of Pest Management
    Summary and Recommendations

    Deciduous Forests 16
    Riparian Woodlands and Forests
    Nonriparian Aspen Woodland
    Deciduous Nonriparian Draws
    Shelterbelts

    Grasslands and Shrubsteppe 22
    Grasslands
    Effects of Livestock Grazing
    Impacts of Pest Management
    Summary and Recommendations
    Shrubsteppe

Habitat Profiles 29

Explanation of Species Accounts 33

Species Accounts 39-183

Species	Page
Turkey Vulture	40
Osprey	41
Northern Harrier	42
Sharp-shinned Hawk	43
Cooper's Hawk	44
Northern Goshawk	45
Broad-winged Hawk	46
Swainson's Hawk	47
Red-tailed Hawk	48
Ferruginous Hawk	49
Golden Eagle	50
American Kestrel	51
Merlin	52
Peregrine Falcon	53
Prairie Falcon	54
Mountain Plover	55
Killdeer	56
Willet	57
Spotted Sandpiper	58
Upland Sandpiper	59
Long-billed Curlew	60
Marbled Godwit	61
Mourning Dove	62
Yellow-billed Cuckoo	63
Black-billed Cuckoo	64
Flammulated Owl	65
Burowing Owl	66
Long-eared Owl	67
Short-eared Owl	68
Common Nighthawk	69
Common Poorwill	70
Chimney Swift	71
Vaux's Swift	72
Black Swift	73
White-throated Swift	74
Ruby-throated Hummingbird	75
Black-chinned Hummingbird	76
Calliope Hummingbird	77
Broad-tailed Hummingbird	78
Rufous Hummingbird	79
Belted Kingfisher	80
Lewis' Woodpecker	81
Yellow-bellied Sapsucker	82
Red-naped Sapsucker	83
Williamson's Sapsucker	84
Olive-sided Flycatcher	85
Western Wood-Pewee	86
Eastern Wood-Pewee	87
Least Flycatcher	88
Hammond's Flycatcher	89
Dusky Flycatcher	90
Willow Flycatcher	91
Cordilleran Flycatcher	92
Eastern Phoebe	93
Say's Phoebe	94
Great Crested Flycatcher	95
Cassin's Kingbird	96
Western Kingbird	97
Eastern Kingbird	98
Horned Lark	99
Purple Martin	100
Tree Swallow	101
Violet-green Swallow	102
Northern Rough-winged Swallow	103
Bank Swallow	104
Cliff Swallow	105
Barn Swallow	106
Brown Creeper	107
Rock Wren	108
House Wren	109
Sedge Wren	110
Marsh Wren	111
Ruby-crowned Kinglet	112
Eastern Bluebird	113
Western Bluebird	114
Mountain Bluebird	115
Townsend's Solitaire	116
Veery	117
Swainson's Thrush	118
Hermit Thrush	119
American Robin	120
Gray Catbird	121
Northern Mockingbird	122
Sage Thrasher	123
American Pipit	124
Sprague's Pipit	125
Cedar Waxwing	126
Loggerhead Shrike	127
Bell's Vireo	128
Solitary Vireo	129
Yellow-throated Vireo	130
Red-eyed Vireo	131
Warbling Vireo	132
Tennessee Warbler	133
Orange-crowned Warbler	134
Nashville Warbler	135
Yellow Warbler	136
Yellow-rumped Warbler	137
Townsend's Warbler	138
Black-and-white Warbler	139
American Redstart	140
Ovenbird	141
Northern Waterthrush	142
MacGillivray's Warbler	143
Common Yellowthroat	144
Wilson's Warbler	145
Yellow-breasted Chat	146
Scarlet Tanager	147
Western Tanager	148
Lazuli Bunting	149
Indigo Bunting	150
Dickcissel	151
Rose-breasted Grosbeak	152
Black-headed Grosbeak	153
Green-tailed Towhee	154
Rufous-sided Towhee	155
Chipping Sparrow	156
Clay-colored Sparrow	157

Brewer's Sparrow	158	McCown's Longspur	172
Vesper Sparrow	159	Chestnut-collared Longspur	173
Lark Sparrow	160	Bobolink	174
Sage Sparrow	161	Red-winged Blackbird	175
Savannah Sparrow	162	Western Meadowlark	176
Lark Bunting	163	Yellow-headed Blackbird	177
Baird's Sparrow	164	Brewer's Blackbird	178
Grasshopper Sparrow	165	Orchard Oriole	179
Fox Sparrow	166	Northern Oriole	180
Song Sparrow	167	Cassin's Finch	181
Lincoln's Sparrow	168	Pine Siskin	182
Swamp Sparrow	169	American Goldfinch	183
White-crowned Sparrow	170		
Dark-eyed Junco	171		

Literature Cited 193

Index 217

Figures:

Figure 1. Public Lands of the Northern Rockies and Great Plains 7
Figure 2. Physiographic Regions of the Northern Rockies and Great Plains 8
Figure 3. Breeding Bird Survey (BBS) Routes in the Northern Rockies and Great Plains ... 8
Figure 4. Coniferous Forest Habitat Profiles 29
Figure 5. Riparian Woodland Habitat Profiles 31
Figure 6. Mixed-Grass Prairie/Shrubsteppe Habitat Profiles 32

Tables:

Table 1. Neotropical Migrants in Taxonomic Order Including Wintering Area 35
Table 2. Neotropical Migrants in Taxonomic Order Organized by Wintering Area 37
Table 3. Avian Species Occurrence in Major Habitat Types of the Northern Rockies
and Great Plains 184

PREFACE

This book covers the geographic region defined by the United States Forest Service as "Region 1" (a.k.a. the "Northern Region")—a vast area extending across the northern tier of states from eastern Washington to North Dakota. The area encompassed includes six states in whole or in part: Montana and North Dakota in their entirety, the northern half of Idaho, and the northernmost portions of Wyoming and South Dakota, as well as the easternmost slice of Washington (Figure 1). The area of coverage is thus defined administratively rather than biologically. The research that I have incorporated into this volume, while specifically germane to the Northern Region, includes work from outside the region where appropriate because biological boundaries exceed the managerial boundaries of the Northern Region. Thus the information in this volume will be directly applicable to areas of the Rockies as far south as southern Colorado/northern New Mexico and to portions of the Great Plains as far south as Colorado and Nebraska.

Fewer than 200 species of landbirds breed within this region, but more than 75 percent of these undertake an annual round-trip migration between their breeding grounds in North America and their wintering grounds that extend south beyond the United States border with Mexico. Our focus is the status, conservation, and management of the 144 species of landbirds that breed within Region 1 and form a subset of species whose populations winter in whole or in part in Mexico, Central America, South America, and the islands of the Caribbean. These areas constitute the "Neotropics," or "New World" tropics (in contrast to the "Paleotropics" or "Old World" tropics of Africa, Asia, and India), hence the initially cryptic phrase "Neotropical migrant landbirds" used to describe the species covered herein.

Why these species? Why this geographic region? Why now? Only recently has it become alarmingly clear that many Neotropical migrants have undergone significant, widespread declines at local and regional levels across at least the eastern portions of their breeding ranges in the United States and Canada. The picture in western North America for this ecologically diverse group of species is far from clear. An assessment of the state of our knowledge about these species at regional levels within the West is a logical beginning from which to embark on efforts to examine population and community patterns of these species and to understand the effects of land-management activities on this diverse group.

Traditionally, management agencies paid virtually no attention to nongame species until populations dwindled to the point of ecological no-return and were declared officially Endangered or Threatened. To continue such a strategy will surely result in failure to preserve overall biological diversity, as a recent assessment of this approach for birds should make clear (Ehrlich, Dobkin, and Wheye, 1992).

In recognition of the need for a proactive rather than reactive response to this potential crisis, the National Fish and Wildlife Foundation proposed the Neotropical Migratory Bird Conservation Program known as Partners in Flight (National Fish and Wildlife Foundation, 1991). This program was conceived as a comprehensive and cooperative effort involving partnerships among federal, state, and local government agencies, private-sector foundations, professional organizations, non-government conservation groups, relevant industries, and the academic communities in both North and Latin America. The primary goal of this venture is to gather together the cooperative and coordinated resources of all concerned organizations and bring them to bear on the improvement of population and habitat monitoring, research, management, education of the public, and the forging of international partnerships. By virtue of their expertise and responsibility for the greatest amounts of potentially critical habitats, the U.S. Fish and Wildlife Service and the U.S. Forest Service are the lead agencies in this effort.

It is my hope that the following pages will provide a useful point of departure for the combined efforts of natural resource managers, conservation organizations, and the concerned public toward the goal of securing the future viability of western North America's Neotropical migrants.

ACKNOWLEDGMENTS

Many people contributed in a variety of ways to making this project a reality. Alan Christensen and the other members of the Neotropical Steering Committee in Region 1 deserve credit for their vision and persistence in conceiving of the idea for a regional handbook devoted to Neotropical migrants. Alan and many of the other committee members (Dan Casey [Montana Department of Fish, Wildlife, and Parks], Steve Gniadek [Glacier National Park], Sally Hejl [Forestry Science Laboratory, Missoula], Dick Hutto [University of Montana], Dave McCleerey [Bureau of Land Management, Missoula], Harvey Nyberg [Montana Department of Fish, Wildlife, and Parks], Chris Paige [Forestry Science Laboratory, Missoula], and Nancy Warren [Flathead National Forest]) graciously reviewed portions of the manuscript and provided useful literature dealing with the region's avifauna. Rolf Koford (Northern Prairie Wildlife Research Center), Terry Rich (Bureau of Land Management, North Dakota/Idaho), and Ken Sanchez (U.S. Fish and Wildlife Service, Bismarck, North Dakota) provided hard-to-find literature and checklists; Dave Genter (Montana Natural Heritage Program) and Craig Groves (Idaho Department of Fish and Game) kindly supplied copies of the latilong distributions for birds of Montana and Idaho, respectively.

Additional reviews of selected species accounts were provided by Dennis Flath (Montana Department of Fish, Wildlife, and Parks), Denver Holt (Owl Research Institute, Inc.), Jeff Marks (University of Montana), Riley McCleland (University of Montana), Terry McEneany (Yellowstone National Park), Fritz Prellwitz (Bowdoin National Wildlife Refuge), Susan Reel (Lolo National Forest), and Phil Wright (University of Montana). James Reichel (Montana Natural Heritage Program) graciously read the entire manuscript and provided a number of useful suggestions.

Singular thanks go to Chris Paige for acting as my primary liaison on the project and for her great patience and competence in responding to my seemingly endless requests for materials from the "gray" literature. I am grateful to Libby Hart of the Rutgers University Camden Library for fulfilling what must have appeared as constantly reiterative requests for papers that invariably were not among the university's library holdings.

I am indebted to Chris Moore for her help in developing and producing her wonderful drawings of birds that enliven the pages of this book. I thank Jack Cooper and Ellen Santasiero of The High Desert Museum's Communications Department for their help in turning my manuscript into a book. Ellen deserves much praise not only for her work on the overall design but also for her execution of the final maps, the habitat profiles, and the cover. I thank her for her patience and her tenacity.

This book is based on a report commissioned by Region 1 of the U.S. Forest Service. Its purpose is to serve as an integral part of efforts aimed at understanding the present population status and conservation needs of the Neotropical migrant landbirds that breed in the northern Rockies and Great Plains. I thank the Forest Service for encouraging me to seek broad dissemination of this work so that it may be read and used widely. I am grateful to Peggy Pace and her staff at the University of Idaho Press for bringing that goal to fruition.

INTRODUCTION

Identification of the Problem: Forest Fragmentation

Of the 646 species of birds that breed regularly in North America, slightly more than half migrate annually to the Neotropics (Ehrlich, Dobkin, and Wheye, 1988). Nearly 260 of these migrants are landbirds, the majority of which depend upon forests and woodlands either for breeding habitat, wintering habitat, or both. Neotropical migrants include most of the species belonging to many familiar groups: warblers, flycatchers, hummingbirds, vireos, thrushes, swallows, and many of the sparrows and other finches.

An alarming and pervasive decline in the North American breeding populations of many previously common and widespread species of birds has become apparent only within the past several years (Terborgh, 1989; Ehrlich, Dobkin, and Wheye, 1992). Consistent patterns of population decline over the past 40 years were demonstrated by a small number of long-term studies of breeding birds in deciduous forests of the eastern United States (reviewed by Askins, Lynch, and Greenberg, 1990). The sites where these studies were conducted typify the characteristics of much of the eastern deciduous forest today—all of the forests were relatively small (<350 ha) and all were spatially isolated from other forested habitats. The population declines documented in these studies confirmed the results of other analyses that found decreased densities or complete loss of many Neotropical migrants in smaller wooded areas when compared to larger forested areas in eastern North America (e.g., Bond, 1957; Forman, Galli, and Leck, 1976; Whitcomb et al., 1981; Ambuel and Temple, 1982; Lynch and Whigham, 1984; Freemark and Merriam, 1986; Askins, Philbrick, and Sugeno, 1987; Robbins, Dawson, and Dowell, 1989). The inescapable conclusion of these and other additional studies is that fragmentation of large, contiguous forests into smaller, isolated patches results in avian communities of predictably different species composition. Larger forests tend to have more species of "forest-interior nesters" and Neotropical migrants, many of which are entirely absent or greatly reduced in number in smaller forests.

By what mechanisms does the process of habitat fragmentation produce the loss of breeding bird densities and loss of species seen in the forest patches? Surveys of nesting success (Gates and Gysel, 1978; Chasko and Gates, 1982; Temple and Cary, 1988; also see Martin, 1992) and experiments with artificial nests (Wilcove, 1985; Yahner and Scott, 1988) indicate that nest predation is higher and nesting success is lower near the forest edge than in the interior of forests. Fragmentation produces more "edge" relative to "interior" for the resulting forest patch, thus leading to increased nest predation. The most common avian and mammalian nest predators (American Crow, Blue Jay, Common Grackle, raccoons, opossums, and domestic cats) all occur in higher densities around forest edges (Bider, 1968; Robbins, 1980; Whitcomb et al., 1981; Rich, Dobkin, and Niles, in press).

In addition to increased susceptibility to predation, Neotropical migrants face another, more subtle threat in fragmented landscapes: the Brown-headed Cowbird. The spread of livestock and European-style agricultural practices in North America over the past 150 years was accompanied by a tremendous expansion of the cowbird's geographic range. The cowbird is an obligate brood parasite that lays its eggs in the nests of other species, leaving the foster parents to rear the cowbird's young, invariably at the expense of the host species' own nestlings. In eastern North America, cowbirds are more abundant in forest edge habitat than in interior forest (Brittingham and Temple, 1983), and brood parasitism by Brown-headed Cowbirds is significantly more frequent near the forest edge (Gates and Gysel,

1978; Brittingham and Temple, 1983; Temple and Cary, 1988; Robinson, 1992).

In the relatively large, contiguous forests of southern New Jersey that are broken only by narrow, forest-dividing corridors, we found that Brown-headed Cowbirds were more abundant than 20 of 21 species of forest-interior Neotropical migrants (Rich, Dobkin, and Niles, in press). Cowbirds were significantly more abundant along corridor/forest edges than in forest interiors, but the absolute difference was very small—cowbirds were distressingly common even in forest interiors! The corridors attract avian nest predators and cowbirds but are not avoided by nesting forest-interior species. These ubiquitous narrow corridors thus create ecological traps for forest-interior Neotropical migrants, and may be important contributors to the widespread declines of forest-interior nesting species in eastern North America.

Forest-interior species and Neotropical migrants in general are highly susceptible to the population impacts produced by increased rates of nest predation and brood parasitism. Compared to resident forest species and short-distance migrants that winter within the United States, Neotropical migrants tend more often to build open cup nests, nest closer to or on the ground, and generally produce only a single, relatively small clutch per season (Greenberg, 1980; Ehrlich, Dobkin, and Wheye, 1988).

It is now apparent that forest fragmentation dramatically alters the resulting avian community. The mechanism driving the disappearance of these species could be losses suffered on the breeding grounds as a result of increased rates of nest predation and brood parasitism experienced by Neotropical migrants in general (and by forest-interior nesters, specifically) in "edge-rich" fragments. The consistent population declines documented in long-term studies conducted in small isolated forest patches, however, are not mirrored in the very few long-term studies conducted in large, contiguous forest tracts (reviewed by Askins, Lynch, and Greenberg, 1990). These results lead to the conclusion that increased predation and brood parasitism may account for the disappearance of these species from smaller forest fragments and for regional declines, without necessarily producing widespread decline throughout a species' entire breeding range.

Tracking Long-Term Population Trends

How then, might changes be detected in avian populations that would signal increases and decreases in numbers throughout a species' breeding range? We have only one source of quantitative information regarding regional and range-wide changes in breeding populations of Neotropical migrants and that is the North American Breeding Bird Survey (BBS). First conducted in 1966 (Robbins, Bystrak, and Geissler, 1986), the BBS is an annual roadside survey of the breeding avifauna in the United States and Canada. BBS data have been used to estimate trends of nearly 370 of the more common species (Robbins, Bystrak, and Geissler, 1986; Droege and Sauer, 1989; Sauer and Droege, 1992). Estimates are based on counts conducted in late May and early June along more than 3,000 established routes, although only two thirds of these are usually run in any given year. Data collection for each route is accomplished by experienced volunteers using standardized procedures to sample avian populations at 50 stops spaced at 0.8 km intervals along secondary roadways. Routes are run beginning one half hour prior to sunrise, and record all birds detected during three-minute periods within a 0.4 km radius at each stop.

What has emerged from analyses of BBS data is a picture of general stability or slight overall increase in eastern populations of many (but not all) Neotropical migrants during the period 1966–78, followed by a general decline in numbers for most of these species from 1979 to the present (Robbins et al., 1989). Most permanent residents and species that migrate within North America ("short-distance migrants") did not exhibit a general pattern of decrease. A recent analysis of the same data using different methodologies arrives at somewhat contradictory conclusions (Böhning-Gaese, Taper, and Brown, 1993), and suggests that nest predation and cowbird parasitism are the more important determinants of population decline in these species. Perhaps the most important point from these studies, however, is that factors contributing to population declines occur both on the breeding grounds in North America and on the tropical wintering grounds.

Unlike, Böhning-Gaese, Taper, and Brown (1993), Robbins et al. (1989) adjusted their data in order to project overall trends for each species (rather than for individual populations) across eastern North America as a whole. Robbins et al. (1989) further analyzed their data to examine species grouped by their use of forested versus nonforested habitats in both breeding and wintering seasons. Their results provide strong evidence that tropical deforestation is indeed contributing to patterns of decline in Neotropical migrants detected on both regional and continental scales. These results clearly argue that Neotropical migrant landbirds of eastern North America are experiencing problems on their tropical wintering grounds that have become increasingly severe since the late 1970s.

Problems in Tropical Wintering Areas

While it is clear that breeding populations are suffering adverse impacts during the breeding season in temperate North America, we have been slow to realize the potential crisis faced by Neotropical migrants on their wintering grounds. The first concerted attempt to assess the biology and distribution of "our" breeding birds on their wintering areas in the Neotropics did not occur until 1977 (Keast and Morton, 1980). That symposium revealed our ignorance not only of the ecological relationships of these species on their tropical wintering grounds but also our lack of knowledge of even such fundamental information as the distribution of many of our most common species in winter. Twelve years later, a similar assessment (Hagan and Johnston, 1992) illustrates how much we have learned in the intervening years, but paints a dire picture of potentially imminent loss of many species' populations, and no less than the complete collapse of the Neotropical migrant system. The problems faced by these birds stem directly from changing patterns of land use, principally habitat destruction and habitat fragmentation.

The relative stability or increase of many forest-bird populations seen in the first half of the BBS coincided with maximum areal coverage by mature forest in the eastern United States relative to the preceding 100 years (Birch and Wharton, 1982; Brooks and Birch, 1988). Although some significant areas exhibited a net loss in mature forest even in recent years (T.W. Birch, cited in Askins, Lynch, and Greenberg, 1990), forest cover overall remained relatively stable when viewed against the tremendous net loss of tropical forest area (Myers, 1980; Melillo et al., 1986; World Resources Institute, 1989) and the accelerating rate of that loss (Fearnside, 1990). The late 1970s (when many Neotropical species commenced their decline in the BBS data) marked the beginning of extensive forest clearing on the Caribbean slope of Mexico and Central America—areas harboring great concentrations of eastern Neotropical migrants (Robbins et al., 1989; Askins, Lynch, and Greenberg, 1990).

The increasing rate of deforestation within the Neotropics is driven by runaway population growth in the Third World and insatiable consumption by the developed world of the tropics' raw materials and agricultural products (Ehrlich and Ehrlich, 1990). The proximal mechanism for forest loss is the development of roads that provides access into previously impenetrable forested areas. Such a pattern is exemplified in Costa Rica, for example, where road development into previously inaccessible areas has accelerated the rate of deforestation from less than 2 percent per year prior to 1977 to more than 7 percent per year beginning in 1977 (Sader and Joyce, 1988). Most alarmingly, tropical deforestation now usually leads to conversion of these previously forested lands to croplands and cattle pasture instead of returning them to forest. The result is a sustained reduction in total forest area.

More than half of the Earth's species are found only in tropical forests (Wilson, 1988). The destruction of these forests and the consequent loss of Earth's biodiversity is not confined to the Neotropics alone; it has become the norm throughout the planet's tropical forests (Gradwohl and Greenberg, 1988). If, however, we focus only on Neotropical migrant landbirds, we can delineate specific geographic areas of greatest concern within the Neotropics. Most of these migrants winter in the areas closest to the United States: Mexico, Guatemala, Belize, Honduras, and the islands of Cuba, the Bahamas, Haiti, and the Dominican Republic. As one proceeds farther south or east, the relative proportion of North American migrants in the tropical avifauna diminishes rapidly. There are, nevertheless, numerous species or populations that winter farther south in Central America, across northern South America, southward through the Andes, and into the Amazon Basin (see Keast and Morton, 1980; Rappole et al., 1983; Hagan and Johnston, 1992).

The concentration of North American migrants into northern Central America and the Greater Antilles illustrates the disparity in vulnerability of Neotropical migrants on their wintering grounds compared with their breeding grounds (Terborgh, 1980, 1989). Some species that breed across the vast land area of Canada and the United States are compressed into wintering grounds less than one fifth the area of their breeding range. Two consequences follow from this reduction in area: (1) extensive deforestation in even limited areas of the wintering grounds will impact disproportionately large numbers of birds, and (2) the effects of negative impacts in wintering areas are distributed over very large geographic areas in the breeding season, thus making it very difficult to detect these wintering impacts by surveying birds on their breeding grounds (Rappole et al., 1983; Wilcove and Terborgh, 1984). Hence, it is all the more remarkable and alarming that we have detected the widespread declines in Neotropical migrant breeding populations that are detailed above.

The Picture in Western North America

As the preceding discussion may have implied, we know far more about the state of Neotropical migrant birds in eastern North America than we do about western populations and species. This is due in large part to the relative paucity of BBS routes in many parts of the West and to the proportionately greater amount of nonurban/suburban habitat in many parts of the West. Given the significant problems of undersampling associated with BBS routes in most parts of western North America, it is disconcerting that a recent extensive analysis of these limited data reveals several sources of concern (Paige, 1990). Paige examined the overall pattern revealed by the 1966–85 BBS data (rather than comparing trends between the early and later periods) and found indications of local or regional declines in 37 species of Neotropical migrants scattered across the West. Although the data indicated that western Neotropical migrants as a group were not declining overall, there was evidence of significant, widespread declines in 19 species of western songbirds (primarily resident species of native grassland and shrubsteppe habitats).

Similarly, insufficient route coverage over much of the West limits the recent attempt to compare trends between periods in BBS data for western populations (Sauer and Droege, 1992). Only 43 of the 144 species of western Neotropical migrant landbirds that breed in Region 1 appeared frequently enough in the database to project meaningful statistical trends. The results overall for the subset of western populations analyzed point to the general absence of declining trends that were seen in eastern populations.

Taken together, the two assessments of western BBS data indicate that western populations of Neotropical migrants as a whole are faring better than eastern North American populations. In seeking reasons for this apparent dichotomy, one important difference in breeding conditions experienced by eastern and western populations is the nature and extent of forest fragmentation. In general, forest fragmentation in the East is a direct result of humanity's activities and produces a fundamentally different landscape when viewed against fragmented forests in the West. The vast expanses of eastern deciduous forest that existed prior to European colonization have been replaced by extensive agricultural landscapes resulting in only small remnant forest stands throughout a significant portion of the East. Additionally, forest fragments are often surrounded by areas converted to suburban and urban landscapes, terrain that, like most of the agricultural areas, will not regenerate into forest second-growth. Wilcove (1990) aptly described the remnants of eastern deciduous forest as "islands of green in a paved or plowed sea."

In contrast, fire and topographic diversity in the West combined in the past to produce a temporally dynamic, naturally fragmented landscape compared with the (previously) extensive and relatively homogeneous eastern deciduous forest. In little more than a century, we have profoundly altered this pattern with fire suppression and extensive timber harvest. Forest fragmentation in the West now results primarily from timber harvest. Nevertheless, landscapes in the West differ fundamentally from the East by virtue of retaining the potential for return to second-growth forest on ecologically meaningful time scales. In addition, old-growth forest in eastern North America is virtually nonexistent, but old growth remains a significant (although extremely reduced and highly fragmented) component of some forested landscapes in the West.

Are western populations and species of forest-dwelling Neotropical migrants affected adversely by forest fragmentation? By virtue of their longer association with naturally fragmented forested landscapes, it may be that as a group they are far less impacted by forest fragmentation on at least some spatial scales in comparison to their eastern counterparts. This question is only beginning to receive attention (Aney, 1984; Rosenberg and Raphael, 1986; Keller and Anderson, 1992; Hejl, in prep.), and we are in desperate need of long-term studies to examine the issue at biologically relevant spatial and temporal scales. No studies have been published that explicitly examine the effects and dynamics of cowbird parasitism and nest predation in the context of fragmented forested landscapes in the West.

Paige's (1990) analysis indicated that in every major habitat type in the West there are key species of Neotropical migrants that warrant our immediate attention and concern (also see Saab and Groves, 1992). Grassland and shrubsteppe habitats in particular may be suffering the greatest problems as demonstrated by widespread declines in their avifaunal communities. As a group, however, relatively few species of Neotropical migrants breed in the arid and semiarid grasslands and shrublands that cover much of the western United States. By comparison, the vast coniferous forests of the West harbor a significant number of Neotropical migrant species, albeit many fewer than occur in eastern deciduous forests. The most species-rich western habitats for Neotropical migrant landbirds are deciduous woodlands and forests, which occur primarily in canyon bottoms and along streams and rivers. Neotropical migrants comprise a greater proportion of the breeding species in arid-land riparian habitats than in any other major western habitat (Bock et al., in press). For example, in montane riparian habitats of central Nevada, Neotropical migrants comprise 85 percent of the riparian-dependent breeding species (Dobkin and Wilcox, 1986).

Unfortunately, western deciduous forests and woodlands are among the most threatened habitats in North America (Terborgh, 1989). More than 90 percent of previously extant desert riparian woodland, extremely important habitat for Neotropical migrants both during breeding and migration, has been destroyed by flood control and irrigation projects (Hendrickson and Kubly, 1984). Riparian woodlands throughout much of western North America have been heavily impacted and frequently degraded by livestock grazing (Bock et al., in press).

The apparent lack of overall population declines among western populations of Neotropical migrants may derive from an essential difference in wintering areas used by eastern and western migrants. It has been argued that western Mexico and Guatemala may be the primary or even exclusive wintering locale for many of the Neotropical migrant landbirds that breed in western North America (Barlow, 1980; Fitzpatrick, 1980; Hutto, 1985). Among the many species wintering there, many utilize disturbed habitats, including a wide range of second-growth forest (Hutto, 1986, 1989), although a significant number also use or are restricted to primary forest (Hutto, 1992). If this difference in wintering areas has buffered western migrants from extensive losses due to conditions encountered on the wintering grounds, the situation may soon change. As in other parts of the tropics, habitat alteration in forests of western Mexico increasingly takes the form of conversion to agriculture and of clear-cutting to produce the early successional stages that are used by very few migrant species (Hutto, 1989).

Like the habitats in which they dwell, our picture of the overall status of these birds is badly fragmented. We know with certainty that widespread declines have occurred in many populations that breed in eastern North America. We also know with equal certainty that these declines are neither limited to the eastern half of the continent nor are they limited only to Neotropical migrant species (Ehrlich, Dobkin, and Wheye, 1992).

For most species, we lack sufficient information to differentiate between the alternatives of relative stability and relative decline among populations of Neotropical migrants breeding in western North America. This presents us both with a problem and an opportunity. In the most optimistic scenario, we have the opportunity to begin now to establish research programs that will evaluate the impacts of our management activities on western populations and to undertake comprehensive, long-term monitoring efforts that will distinguish patterns of population change. To the extent that conditions on the breeding grounds control population trends of these species, actions that we undertake now offer the possibility of preventing or at least mitigating the pattern of losses seen among eastern populations.

Searching for generalizations about population change in "Neotropical migrant landbirds" as a group will prove fruitless, in large part because these species vary greatly in their life histories and in their seasonal distributions (Ehrlich, Dobkin, and Wheye, 1988; Sauer and Droege, 1992). Success in our efforts will likely be achieved only by combining a focus at the species level with a habitat-based conservation approach. There is, clearly, much work to be done.

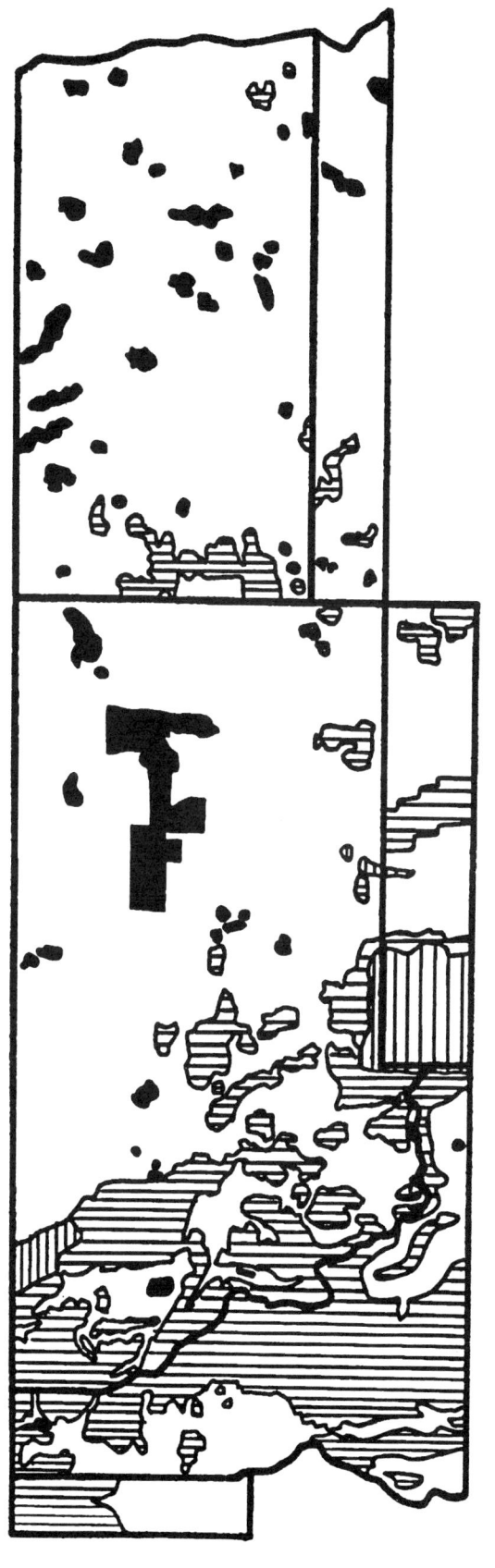

Figure 1. United States Forest Service (▭), Fish and Wildlife Service (■), and National Park Service (▭), lands of the Northern Rockies and Great Plains. Additional areas administered by the federal Bureau of Land Management and by various state agencies occur extensively throughout the Northern Region but generally are too small and discontinuous to be depicted here.

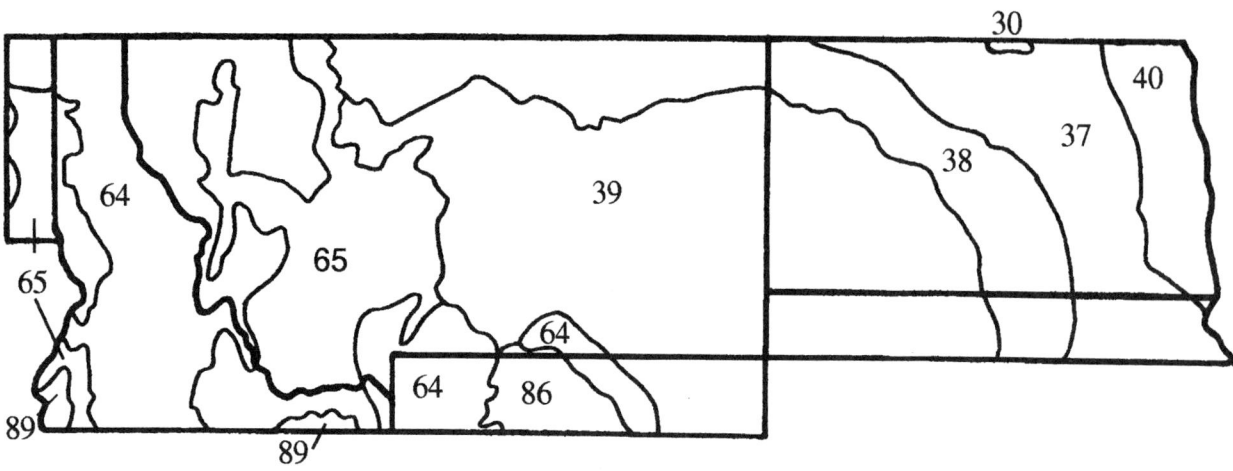

Figure 2. Physiographic regions of the Northern Rockies and Great Plains (based on Robbins, Bystrak, and Geissler, 1986): (30) Aspen parklands, (37) Drift prairie, (38) Missouri Plateau-glaciated, (39) Missouri Plateau-unglaciated, (40) Black prairie, (64) Central Rocky Mountains, (65) Dissected Rocky Mountains, (86) Wyoming Basin, (89) Columbia Plateau.

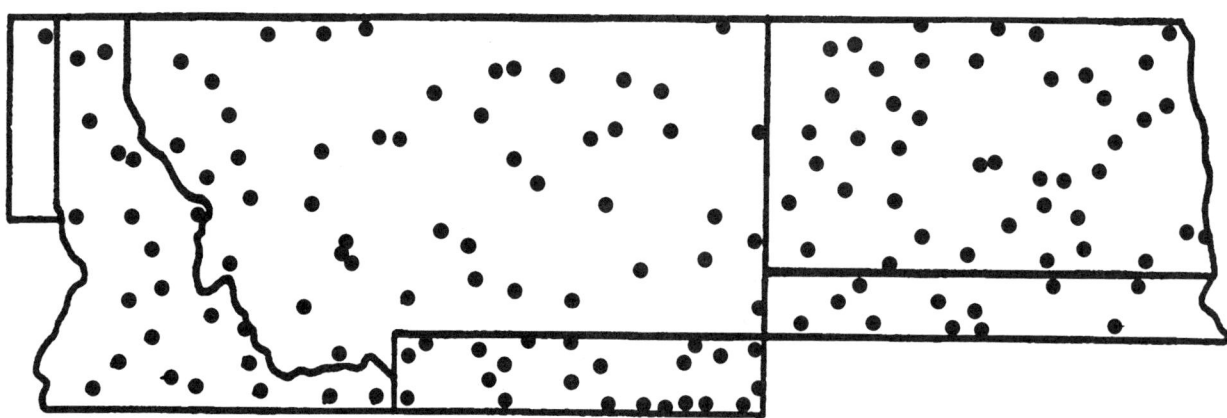

Figure 3. Breeding Bird Survey (BBS) routes of the United States Fish & Wildlife Service in the Northern Region.

SURVEY OF MANAGEMENT ACTIVITIES AND IMPACTS IN THE NORTHERN REGION

Coniferous Forests

It is both enlightening and distressing to peruse the workshop proceedings entitled "Management of Western Forests and Grasslands for Nongame Birds," published in 1980 by the U.S. Forest Service (DeGraff and Tilghman, 1980). There is much of relevance, and many of the warnings and recommendations contained therein, especially regarding coniferous forests of the Northern Region and the Pacific Northwest (e.g., Sanderson et al.; Smith; Miller and Miller; Mannan) are still pointedly relevant today for the region's Neotropical migrant landbirds. In the 10 years following publication of the proceedings, the pace and extent of timber harvest in the coniferous forests of the Northern Region and the Pacific Northwest achieved unprecedented levels. Most dramatic has been the reduction in areal extent of virtually all types of old-growth conifer forests and the extensive use of clear-cutting in timber harvest.

Any attempt to design a comprehensive conservation and management strategy for Neotropical migrants in conifer-dominated landscapes is inexorably confronted by the huge gaps in our knowledge of specific habitat use, nesting and foraging requirements, survival rates, and population viability for the great majority of species. Overlying this general lack of specific information is the additional set of questions addressing how these parameters vary among different silvicultural treatments (Finch, 1991b).

Variation in topography and past fire frequency creates an inherently fragmented landscape in montane coniferous-forest ecosystems. This intrinsic quality of the landscape has been augmented greatly by stand-level management impacts (clear-cutting, selection harvest, thinning, fire suppression, and fire prescription) superimposed on the natural variation. Couple this landscape-scale heterogeneity with highly variable fluctuations in climate between and within years (e.g., Franzreb and Ohmart, 1978; Szaro and Balda, 1979; Smith, 1982; Morrison et al., 1987; Hejl, Verner, and Balda, 1988) and the likelihood becomes vanishingly small for achieving commensurate results among short-term studies of avian communities.

We are only beginning to address the sorts of questions that have been examined in the East, and to assess the kinds of factors that have been implicated as important in population declines of Neotropical migrants inhabiting eastern forest habitats. The explicit question of the impacts of forest fragmentation on birds in western coniferous forests is addressed in only four studies to date (Aney, 1984; Rosenberg and Raphael, 1986; Keller and Anderson, 1992; Hejl, in prep.). Studies in the central Sierra Nevada by Jerry Verner and associates (Hejl, Verner, and Balda, 1988; Verner and Larson, 1989) compared avian communities and vegetation parameters at many sites over a significant geographic area and across silvicultural treatments, but have not yet explicitly incorporated a landscape-level consideration of forest distributions in the analyses.

No studies have been conducted to examine the relationship between cowbird nest parasitism and forest fragmentation in the West. Martin (1988) has made a significant start in evaluating nest predation with his work in small forest fragments of ponderosa pine in central Arizona. His results point to the potential importance of nest predation in fragmented western forests, but many more studies are needed in other types of coniferous forest and in other regions. In Douglas fir/ponderosa pine forests of the northern Rockies, Hejl (pers. comm.) found a highly significant association between cowbird abundance and the amount of nonforested (i.e., clear-cut) land at the landscape scale. These results are potentially ominous for forest-nesting songbirds and point to the dire need for data to determine whether patterns of nest predation and nest parasitism in western coniferous forests will parallel the effects seen in eastern deciduous forest fragments.

Although a number of recent and current studies in the Northern Region address the question of species distributions among different seral stages and silvicultural treatments, only one (Hejl, in prep.) is framed explicitly and quantitatively in the context of landscape-scale fragmentation to address the role of fragmentation in the patterns being documented. It may well be that in the Northern Region we will never know the full effects of such fragmentation on old-growth, Douglas fir/ponderosa pine ecosystems, as only relatively small remnant patches remain in the northern Rockies (Aney, 1984; Hejl and Woods, 1991; Moore, 1992).

Douglas Fir and Ponderosa Pine

As the most economically important and geographically widespread forest types in western North America, Douglas fir (*Pseudotsuga menziesii*) and ponderosa pine (*Pinus ponderosa*) forests have received the most attention. In order to provide meaningful data, studies must: (1) cover a sufficiently large geographic area to minimize the statistical effects of site-specific peculiarities, (2) include adequate replication across different treatments (silvicultural treatments, stand age, stand size, etc.), including pretreatment sampling, and (3) extend over enough years to incorporate (and hopefully distinguish) the "background noise" of climatically induced variation in populations. The one additional (and most challenging) feature required of these studies is the explicit incorporation of a landscape context into their design and analysis. To date, only a few studies anywhere in western North America encompass these attributes; in the Northern Region, only Hejl's work (in prep.) explicitly incorporates adequate landscape-scale considerations.

In northwestern California, Rosenberg and Raphael (1986) specifically investigated the patterns of species distributions and abundances in relation to forest patch size, degree of insularity, and edge characteristics in Douglas fir forests. Their study encompassed 46 stands ranging in size from 5 to more than 300 ha. The results contrast markedly with similar analyses of eastern deciduous forest avifaunas. Among the subset of species that were sufficiently abundant to be analyzed statistically, the species absent from smaller fragments (i.e., those exhibiting significant area effects) were mostly resident species rather than migrants. Among Neotropical migrant species, only the Sharp-shinned Hawk exhibited such an effect. Unfortunately, only 44 of the 101 species of birds found in the study occurred frequently enough to be analyzed statistically. Hence, for the majority of species we know nothing about the potential influence of fragmentation. The relevance of this study for the Northern Region is uncertain because of substantial climatic and faunal differences from interior Douglas fir forests.

Aney (1984) surveyed old-growth ponderosa pine/Douglas fir forest patches in western Montana to examine the relationship between forest remnant size and avian community composition. He found a significant correlation between species richness and forest size, thus paralleling the results seen in remnant forest patches of eastern deciduous forest. Among the Neotropical migrants that appeared to exhibit significant area effects were Sharp-shinned and Cooper's Hawks, Williamson's Sapsucker, *Empidonax* flycatchers, Brown Creeper, and Solitary Vireo. The results of Aney's (1984) study are limited by the strongly skewed sample of available patch sizes (14 of the 19 stands were smaller than 7 ha, and only one stand larger than 40 ha could be located), and by having been conducted over only a single field season. In spite of these important shortcomings, this study points to the potential for adverse impacts on Neotropical migrants in the Northern Region that result from old-growth fragmentation.

Preliminary results have been published (Hejl and Woods, 1991) for the first year of a three-year study designed to compare avifaunas in remnant old-growth Douglas fir/ponderosa pine forests and older rotation-age forests in western Montana and adjacent Idaho. Although species richness was similar among sites, a greater number of species was found in old growth overall (14 of 68 species were found only in old growth; 7 species occurred exclusively in rotation-age stands). The survey of 16 old-growth (200+ yrs) and 16 rotation-age (80–120 yrs—the oldest managed stands under current prescriptions) stands revealed strong, site-specific differences—emphasizing the danger of drawing inferences from studies encompassing only a few sites. Although results from this study generally agreed well with those of Mannan and Meslow (1984), disparities between them (and Aney, 1984) likely reflect differences among years (each study analyzed data from only a single year).

Several species of Neotropical migrants exhibited significantly greater abundances in old growth (Hejl and Woods, 1991; Hejl, in prep.), including Hammond's Flycatcher, Swainson's and Hermit Thrushes, Townsend's and Yellow-rumped Warblers, and Western Tanager. (Northern Flicker and Pileated Woodpeckers also were significantly more abundant in old growth.) Northern Goshawk and Winter Wren occurred only in the old-growth stands. Only three Neotropical migrants were significantly more abundant in rotation-age stands: Dusky Flycatcher, Solitary Vireo, and Chipping Sparrow. Brown-headed Cowbirds also were significantly more abundant in the rotation-age stands. Four Neotropical migrants typically associated with earlier seral stages were absent from old growth but present in rotation-age stands: Mourning Dove, Cedar Waxwing, Western Meadowlark, and Vesper Sparrow. Within the old-growth stands, the presence of several species was attributed in large part to the presence of large snags—key ecological features of old-growth forest that were entirely absent from rotation-age stands.

A recently completed four-year study in Douglas fir forests of southwestern Montana (Moore, 1992) compared avian communities among old-growth (four stands [70–300+ yrs]), rotation-age (four mature stands [65–145 yrs]), shelterwood cut (one stand [65–145 yrs]), and clear-cut (one stand) stands. The landscape context was typical of the region—extensive fragmentation resulting from the dispersion pattern of harvest units. As in the study by Hejl and Woods (1991), no difference was found in species richness between old and mature stands, but there were significant differences among species. Not surprisingly, the shelterwood (selection cut) provided postharvest habitat with greater abundances and species richness than the clearcut.

The importance of the landscape-level scale of heavy fragmentation is epitomized by the relatively greater abundance found for cowbirds in the old-growth stands compared with the other treatments! The degree of fragmentation in this study likely obscures patterns for some individual species and creates misleading patterns for others. For example, although Williamson's Sapsucker was associated with old-growth stands having large-diameter snags, and Hammond's Flycatcher was significantly more abundant in old growth, both American Robin and Chipping Sparrow also were significantly more abundant in old growth. In fact, Moore (1992) documents more similarities than differences between old-growth and mature avifaunas (both of which contrasted strongly with harvest units), most likely due in large part to small stand sizes embedded within a heavily fragmented landscape; Mountain Chickadee, Red-breasted Nuthatch, Brown Creeper, Ruby-crowned Kinglet, Swainson's and Hermit Thrushes, Warbling Vireo, Yellow-rumped Warbler, and Western Tanager did not differ in relative abundance between old-growth and mature stands, but all were significantly more abundant there than in the harvest units. Not surprisingly, White-crowned Sparrow and Cassin's Finch were significantly more abundant in the harvest units.

In similar old-growth Douglas fir/ponderosa pine habitat in northeastern Oregon,

Mannan and Meslow (1984) compared avian communities in four old-growth stands (200+ yrs) and four rotation-age stands (85 yrs—the oldest managed forest stands in the region). In their three-year study, the presence of large snags in old growth and absence in rotation-age stands translated into differences in presence and abundance for hole-nesting species, most markedly for Pileated Woodpecker, Vaux's Swift, and sapsuckers. Not surprisingly, ground-foraging species were better represented in the rotation-age stands. Northern Goshawk, Flammulated Owl, Red-naped and Williamson's Sapsuckers were found only in the old-growth stands; Mourning Dove and Calliope Hummingbird were found only in the rotation-age stands.

Mannan and Meslow (1984) found significantly greater abundances in old growth for Brown Creeper, Red-breasted Nuthatch, probably Hammond's Flycatcher (identification was apparently confounded with Dusky Flycatcher), Golden-crowned Kinglet, Swainson's, Varied, and Hermit Thrushes, and Townsend's and MacGillivray's Warblers. In rotation-age stands, Dusky Flycatcher, Ruby-crowned Kinglet, Chipping Sparrow, Dark-eyed Junco, Brown-headed Cowbird, and Cassin's Finch were all significantly more abundant.

Studies conducted in ponderosa pine forests of the southern Rockies similarly address the question of silvicultural treatment on avian community structure (Szaro and Balda, 1979; Brawn and Balda, 1988), although there are significant differences in avian species composition compared with ponderosa pine forests in the Northern Region. Analyses focus on the great variation in avian populations among years due in part to variable weather patterns (Gaud, Balda, and Brawn, 1986; Szaro and Balda, 1986; also see Hejl, Verner, and Balda, 1988) and to variation in avian foraging behavior among years (Szaro, Brawn, and Balda, 1990). As elsewhere, the importance of snag retention in ponderosa pine forests of the southern Rockies has been emphasized for a number of Neotropical migrant species (Balda, 1975; Cunningham, Balda, and Gaud, 1980).

A variety of additional studies comparing avian communities among contrasting silvicultural treatments in various types of coniferous forest offer little insight into the questions they attempt to address (e.g., Ramsden, Lyon, and Halvorson, 1979; Peterson, 1982; Medin, 1985; Medin and Booth, 1989). These studies suffer from lack of replication among treatments (some sample only a single site per treatment or provide "pseudoreplication" by simply subdividing a single site into two or more plots) or from lack of temporal perspective (sampling was conducted in only a single season and site-specific pretreatment sampling was generally lacking), or both; several studies that sampled across years are compromised by statistical analyses that fail to properly assess between-year variation. Prior to 1980, the importance of forest patch size in a landscape context simply was not considered in most studies of avian communities in western coniferous forests (but see Thompson, 1978).

Other Coniferous Forests

Although our knowledge of avifaunas in Douglas fir and ponderosa pine forests in the Northern Region is at best only fragmentary, we know even less about the dynamics of avifaunas in other types of forests.

Keller and Anderson (1992) explicitly examined the effects of fragmentation on birds in subalpine forests of southeastern Wyoming. Only 16 species of breeding birds were detected in unfragmented and fragmented stands of Engelmann spruce (*Picea engelmannii*) and subalpine fir (*Abies lasiocarpa*). During the two years of the study, a marked decrease in overall avian numbers occurred between years, which was attributed to climatic factors. Site-specific differences appeared to be more important than fragmentation effects in affecting species distributions between treatments, with the exception of Brown Creepers, which never occurred in fragmented stands. Given the comparatively harsh environmental conditions experienced at high elevations, the results of this study are perhaps not surprising. Significant differences in abundance did occur between treatments: among Neotropical migrants, Yellow-rumped Warbler, Hermit Thrush, American Robin, (and Brown Creeper) were significantly more abundant in unfragmented stands than in fragmented forests. Only Pine Siskins were more abundant in fragmented stands. Overall species composition in stands interrupted by small clearcuts was similar to the composition in unfragmented forests, aside from the loss of Brown Creepers.

Keller and Anderson (1992) argue that the loss of resources by clear-cutting in subalpine forests may convert already marginal habitats into wholly unsuitable nesting areas. Their results are consistent with those of Franzreb and Ohmart (1978) and Mannan and Meslow (1984) for Brown Creepers, Hermit Thrushes, and Red-breasted Nuthatches, suggesting that fragmentation in these subalpine forests produces avifaunas that are more similar to managed stands than to unlogged old growth.

In the Northern Region, the most significant study to date outside of Douglas fir/ponderosa pine forest was conducted in old-growth western larch (*Larix occidentalis*)/Douglas fir in northwestern Montana (Tobalske, Shearer, and Hutto, 1991). Within a heavily fragmented landscape, they compared avian communities in four small clearcuts, three small partial cuts, an unlogged forest fragment of 134 ha, and an unlogged research natural area of 339 ha. All snags and all deciduous trees (birch, cottonwood, and aspen) were retained in clearcuts and partial cuts. Although 10 replicates were assessed within each treatment, many species (19 of 51 species) occurred too rarely to analyze statistically.

Golden-crowned Kinglet, Swainson's and Varied Thrushes, and Townsend's Warbler were significantly more abundant in unlogged forest than in the cutting units (Tobalske, Shearer, and Hutto, 1991). Only Tree Swallow, Dark-eyed Junco, and Pine Siskin were significantly more abundant on partial and clear-cut sites compared with the two unlogged sites combined. However, the importance of landscape-scale fragmentation can be seen in the striking similarities found in abundances of several species (e.g., Golden-crowned and Ruby-crowned Kinglets, American Robin, Varied Thrush, Townsend's Warbler, Chipping and Fox Sparrows) in partial-cut sites and surrounding (i.e., fragmented) unlogged forest when contrasted with the larger, more intact, unlogged forest on the research natural area. Golden-crowned Kinglets were far more abundant than Ruby-crowns in the research natural area but not in the unlogged fragment, and only Ruby-crowned Kinglets occurred in the clear-cut treatments. These results are consistent with other studies indicating that Golden-crowned Kinglets are more sensitive to forest fragmentation in the Northern Region than are their congeners.

Echoing the earlier work of McClelland and Frissell (1975), Tobalske, Shearer, and Hutto (1991) emphasize the importance of retaining conifer snags and all deciduous trees (both live and dead) in cutting units. They also found (as did Franzreb [1977] for House Wren and Dark-eyed Junco) that the retention of logging slash supplied critical cover, foraging, and nesting sites for some species. Tobalske, Shearer, and Hutto (1991) provide a cogent discussion of the role of fragmentation and the lack of sufficient replication in decreasing the ability of studies such as theirs to fully address the effects of silvicultural treatments on these avian communities. These limitations result, in part, from the great annual variation in populations of some species (such as Pine Siskin) and the numerical rarity of many species. In their study, rarity precluded analysis of nearly 40 percent of the detected species.

Avian species composition in relation to forest age also has been examined in Douglas fir/western hemlock (*Tsuga heterophylla*) forests on the western slope of the Cascades (Manuwal and Huff, 1987 and references cited therein), but the presence of important component species that do not occur in the Northern Region severely complicates any attempt to extrapolate their results to the northern Rockies.

Fire and Avian Community Response

Few detailed studies have been conducted to explore the effects of fire on birds in western coniferous forests. Most are anecdotal and short term, having been conceived opportunistically in response to forest fires. There is a critical need for well-planned, statistically viable studies of avian community response to fire in all the region's coniferous forest ecosystems. Of even greater urgency is the need for research designed to address the impact of postfire salvage logging on birds that nest and forage in these postfire habitats. Such studies are entirely lacking to date.

In one of the better designed analyses, Skinner (1989) examined several postfire conditions in lodgepole pine *(Pinus contorta)* forests and ecotones in northwestern Wyoming. Within a heavily mosaicized landscape, she examined 22 plots for passerines (also hummingbirds and woodpeckers) in lodgepole/sagebrush and lodgepole/riparian ecotones in a single season. Her results typify the complexity of seral stage avifaunal dynamics. Shrub- and ground-foraging species were significantly more abundant and diverse in two-year postfire ecotones compared with unburned ecotones. Aerial foragers and salliers were more abundant in both two-year and six-year postfire ecotones compared with unburned ecotones. Within these broad guild designations, many species did not conform to the preceding generalities, leading Skinner to emphasize the need to examine species on an individual basis rather than combining species into guilds.

In a recently completed study (Hutto, in prep.; pers. comm.), the effects of intense forest fires in the Northern Region during 1988 were surveyed across 38 different mixed-conifer forests in the two following years. Many avian species that are well known as early postfire associates (e.g., Black-backed Woodpecker) were virtually restricted to these burned forests. Most surprisingly, several Neotropical migrant species (including Olive-sided Flycatcher, Mountain Bluebird, and Chipping Sparrow) were found to occur significantly more frequently in early postfire habitats than in any other forested habitats in the northern Rockies. For Mountain Bluebirds in particular, these habitats appear to be critical for nesting populations. This study clearly indicates the tremendous potential for adverse impact on many avian species that could result inadvertently from postfire salvage logging.

Bock and Bock (1983) analyzed the effects of prescribed burns in ponderosa pine forests and pine-grassland savannah in the Black Hills of South Dakota. In the first and second years following fire, species composition did not differ between burned and unburned control plots. Significant differences in relative abundances were seen in seven Neotropical migrants (American Robin, Mountain Bluebird, Solitary Vireo, Yellow-rumped Warbler, Western Tanager, Dark-eyed Junco, and Chipping Sparrow), which were more abundant on burned plots than on control plots in the first year postfire, although in the second year only the Vesper Sparrow was more abundant on burned plots. In general, their results are similar to those of other earlier studies (cited by Bock and Bock, 1983) in suggesting that the significantly increased abundances seen in the initial postfire breeding season were likely due to an increased food supply resulting from burning.

In western Montana, Lyon and Marzluff (1985) attempted to compare avifaunas in unlogged Douglas fir forest and similar forest that experienced an uncontrolled burn. They examined two sites in the year of the fire and again two years postfire. Little change in total numbers of birds or in species richness was found, but no detailed analysis was performed at the species level. Their study is further complicated by postfire salvage logging and by postbreeding season movements of birds into the burned areas, attracted perhaps as a result of increased foraging opportunities presented by elevated insect densities.

Effects of Livestock Grazing

Virtually nothing is known about the potential influence of livestock on Neotropical migrants in western coniferous forests. We know only that no native ungulate herds in any way comparable to sheep or cattle coevolved with the vegetation of these forests. Hence, we should expect that domestic livestock have exerted a considerable effect on these ecosystems, in terms of vegetation composition and structure (Bock et al., in press). Birds most likely to be affected negatively by livestock grazing in montane forests are species (such as Nashville Warbler, Fox Sparrow, and Lincoln's Sparrow) that are dependent on herbaceous and shrubby ground cover for nesting or foraging.

Studies are needed both during nesting and migration periods to examine the long-term and short-term effects of grazing. Comparisons should be conducted among replicated forested stands with known differences in grazing regimes or grazing histories. Quantitative assessment of vegetation structure and composition must be an integral part of such studies.

Impacts of Pest Management

The control of insect pest outbreaks with the use of chemical control agents is a common silvicultural practice. Many Neotropical migrants inhabiting conifer forests feed almost exclusively on lepidopteran larvae and adults during much of the breeding season and rely heavily on these insects for feeding their nestlings (Ehrlich, Dobkin, and Wheye, 1988; Holmes, 1990). Several species of wood warblers and other Neotropical migrants are well known as specialists on cyclical "outbreak" types of forest insects, and their populations closely track these outbreaks (see e.g., Morse, 1989). Similar relationships likely exist among some species nesting in western coniferous forests, but this question is unexplored.

Impressive quantities of insect pests at all life stages (eggs, larvae, pupae, and adults) are routinely consumed by forest birds (e.g., Torgersen, Mason, and Campbell, 1990), but this fact appears to be largely unappreciated by forest managers. Insectivorous birds, foliage-foraging ants, wasps, and some small mammals are major predators of insect pests in temperate forests. The general effectiveness of birds as control agents of forest insects is often underestimated by focusing on their impact at the height of insect outbreaks. Birds in temperate forests appear to depress or maintain insect numbers at low levels and probably extend the number of years between outbreaks in irruptive pest species (reviewed by Holmes, 1990).

Insectivorous birds and foliage-foraging ants both respond positively to the presence of standing and downed dead trees (Torgersen, Mason, and Campbell, 1990). Hence, retention and recruitment of snags will enhance populations of these natural enemies and augment their effectiveness as natural control agents of insect pests. For many insectivorous birds, the potentially devastating impact on food supplies that presumably results from pest control spraying is a topic in need of study.

The use of selective herbicides to control growth of deciduous trees and shrubs in managed conifer forests is another topic that is unexplored in terms of its impact (through the elimination of required nesting and foraging cover) on Neotropical migrants. A secondary impact of these control measures is the reduction of food resources (insects) for insectivorous birds by virtue of the diminished habitat that supports many insects.

Summary and Recommendations

Avian communities vary in structure and composition, and our perception of these communities is often determined by the temporal and spatial scales we use to delimit them (Wiens, 1989a, b). Consequently, a study-by-study assessment of results for Neotropical migrant species yields sometimes contradictory information and can thoroughly confound attempts to generalize at the species level. In spite of these problems, several salient points for natural resource managers should be noted.

In clear-cut harvesting, the importance of retaining snags, standing live trees, substantial slash, and live deciduous trees (if present) has been emphasized repeatedly (Balda, 1975; Franzreb, 1977; Cunningham, Balda, and Gaud, 1980; Miller and Miller, 1980; Bull et al., 1980; Tobalske et al., 1991; Moore, 1992). Many studies provide clear evidence that it is far more beneficial to retain larger snags rather than smaller ones (McClelland et al., 1979; Mannan and Meslow, 1984; Hejl and Woods, 1991).

We know very little about the impacts of forest management activities on the productivity of Neotropical migrants. Wherever possible, we should move beyond collecting data that simply indicate presence or absence of species and begin to more directly address real population parameters by measuring population productivity (nesting and fledging success) across different management treatments. Attempts to deduce the health and viability of populations based solely on surveys of the number of territorial birds will mask the underlying reality of population sinks in which reproduction is so low that it fails to compensate for mortality (e.g., Gibbs and Faaborg, 1990; Robinson, 1992). Intensive single-species studies that measure productivity across different management treatments will provide the greatest insight into the effects of forestry practices on breeding populations of Neotropical migrants.

Although species may sometimes be conveniently grouped into functional ecological guilds based on shared ecological similarities (nesting location, foraging mode, etc.), they must not be viewed simply as interchangeable units. Each species requires individual assessment for its response to management activities in forest ecosystems, and geographic variation in these responses among populations within species should be expected as the norm, not the exception. To focus on community-wide measures (such as avian species richness or species diversity) for assessment of management impacts is to miss the point of species-specific patterns and responses. Replacement of species that are sensitive to habitat fragmentation or associated with rare or diminishing habitats by species that thrive in human-altered landscapes is not a desirable outcome—even if overall species richness or diversity remains the same.

In designing research projects to address the conservation and management of Neotropical migrants in western coniferous forests, the following elements are essential:

1) It is imperative that we incorporate a landscape-level scale of habitat variation into our research design and our management plans.
2) Long-term studies of at least 5 to 10 years' duration are needed in order to encompass even a minimally reasonable range of background fluctuations driven by climatic variation.
3) Sampling must cover a sufficiently large geographic area to minimize the statistical effects of variation resulting from site-specific differences.
4) Adequate replication across different treatments must be an integral part of any study and ideally should include pretreatment sampling.

Deciduous Forests

The most widespread deciduous forests in the Northern Region are riparian woodlands, which vary from extensive floodplain forests associated with large rivers on the plains to narrow bands of aspen *(Populus tremuloides)* woodlands and willow *(Salix* sp.) thickets along small streams in the mountains. The other major types of deciduous forests in the region are montane aspen forests in the Rockies, wooded draws in badland areas of the westernmost plains, and planted shelterbelts scattered throughout the northern plains.

As is true for most birds in western coniferous forests, our knowledge of species-specific habitat use, nesting and foraging requirements, survival rates, and population dynamics is quite fragmentary for most species associated with western deciduous forests. Overlying the general lack of specific information is the additional set of questions addressing how avian life-history parameters and ecological relationships vary among different management activities, principally livestock grazing and timber harvest.

Riparian Woodlands and Forests

Although constituting less than 1 percent of western landscapes, riparian forests and woodlands harbor the most species-rich avifaunas of all the major habitats found in the western United States (Knopf et al., 1988). Neotropical migrants usually comprise the majority of species in these riparian communities. For example, 60 percent of Idaho's Neotropical migrant landbirds are associated with riparian habitats (Saab and Groves, 1992), and 82 percent of all nesting species in northern Colorado use riparian areas (78 percent of Colorado's landbird species are Neotropical migrants; Knopf, 1985). In montane riparian communities in central Nevada, 85 percent of the riparian-dependent breeding species are Neotropical migrants (Dobkin and Wilcox, 1986).

Unfortunately, western riparian ecosystems are focal points of maximum potential conflict among competing users for livestock grazing, timber harvest, recreation, channelization for flood control, and water diversion for agricultural use, power generation, and domestic consumption (Thomas, Maser, and Rodiek, 1979). In the western United States, the negative effects of human impacts on Neotropical migrants are likely to be most dramatic in these riparian woodlands, which support the greatest diversity and abundance of Neotropical migrants in the West, both in the breeding season and during migration.

Aside from the sometimes expansive floodplain forests in the eastern portion of the region, riparian habitats in the Northern Region tend to be narrow and highly linear. The effects of forest fragmentation seen in eastern deciduous forests therefore might be expected to be much less important for these avifaunas, and the question is virtually unexplored. The only study to address the effects of forest fragmentation on birds in western riparian areas was conducted in montane riparian habitats of central Nevada (Dobkin and Wilcox, 1986). Their results clearly indicate that most riparian species exhibit area-dependent distributions, and that many species are lost from smaller riparian fragments. No assessment of forest fragmentation on avian communities has been conducted in the more expansive floodplain forests of the northern Great Plains where the negative effects of fragmentation should most closely parallel the experience of birds in eastern deciduous forests.

There are no long-term studies that characterize avian communities and their population fluctuations in relatively undisturbed riparian habitats of the Northern Region. Such studies are sorely needed to serve as baselines against which comparisons can be made with riparian communities subjected to various management activities. Especially in the northern plains, we have only "snapshot" qualitative pictures of these rich communities (e.g., Kroodsma, 1973; Tubbs, 1980; Hopkins, Cassel, and Bjugstad, 1986; Douglas et al., 1992). Multiyear studies encompassing a broad range of intact forest sizes and spanning a broad geographical distribution are urgently needed. Equally important is the need for detailed assessments of the use of riparian areas in the region by birds during migration. As critical as these habitats are for breeding populations, they are of paramount importance as migration corridors that provide cover and food, especially during autumn migration. Field studies are desperately needed to quantify these relationships.

Most of the work examining riparian avifaunas has been in the context of livestock grazing impacts on riparian vegetation and the consequent implications for breeding birds. In spite of the contentiousness of this issue, only a handful of studies have been conducted that compare different grazing regimens or histories, and these are reviewed in detail by Bock et al. (in press). Many of these studies suffer incurably from the combination of short-term perspective (i.e., conducted for only a single season) and complete lack of replication among treatments (e.g., Medin and Clary 1990, 1991). Little insight can be gained from such narrowly focused and poorly designed studies (Szaro, 1991). In spite of the general paucity of adequate research, a number of consistent generalities regarding riparian avifaunas can be drawn from the small number of well-executed studies.

The presence or absence of many Neotropical migrant species in riparian habitats is intimately tied to the complexity and density of vegetation structure, especially in the shrub and herbaceous layers. Cottonwood-dominated habitats at lower elevations tend to have the greatest layering of vegetation and support the richest avifaunas (Knopf, 1985; Finch, 1989a, b). In riparian zones, the most obvious effect of livestock grazing on vegetation is to remove (to varying degrees depending on the timing, duration, and number of livestock) the lower vegetation layers. This effect is compounded by continued grazing, which prevents recruitment by woody species and leads to decadence and senescence of the trees and shrubs. The effects of prolonged use by cattle in most riparian habitats of the western U.S. are not subtle and can be easily seen. Such impacted riparian areas support plant and animal communities that are both structurally and taxonomically impoverished relative to intact riparian habitats. Hence, it should come as no surprise that many Neotropical migrants are strongly affected by livestock-induced changes in riparian vegetation (Bock et al., in press).

The lack of recruitment of young trees in cottonwood-dominated floodplain forests is an increasingly common phenomenon that results singly or in combination from altered stream flows, intense livestock grazing, and invasion of exotic woody plant species, especially Russian-olive *(Elaeagnus angustifolia)* (Olson and Knopf, 1986; Knopf and Scott, 1990). The lack of tree regeneration within aging cottonwood forests will inexorably lead to the loss of large snags and live trees without replacement, resulting in significant declines of cavity-nesting species (Sedgwick and Knopf, 1990).

Birds that are most directly and negatively affected by livestock in riparian areas are species that nest or forage in dense shrub or herbaceous ground layers (Bock et al., in press). Among these species are Willow Flycatcher, Wilson's Warbler, Lincoln's and White-crowned Sparrows in the central Rockies (Knopf, Sedgwick, and Cannon, 1988; Schulz and Leininger, 1991), Common Yellowthroat and Yellow-breasted Chat in northeastern Colorado (Sedgwick and Knopf, 1987), Nashville and MacGillivray's Warblers, Common Yellowthroat, and Lazuli Bunting in western Montana (Mosconi and Hutto, 1982). Veery, Savannah and Fox Sparrows, and many others could be added to this list but adequate studies are still lacking.

Some species that require open areas for foraging do respond positively to livestock grazing in riparian habitats (e.g., Killdeer, American Robin, and Brewer's Blackbird), but these are generally species that are widespread and occur more commonly in other types of habitats, or abundant species that are broadly distributed in human-altered habitats.

An important and largely unexplored ancillary impact of cattle in riparian areas results from their relationship to Brown-headed Cowbirds. Several studies now in progress (such as J. Sedgwick's work on Willow Flycatchers, pers. comm.) are finding unusually high incidences of cowbird nest parasitism associated with the presence of cattle in riparian areas. The cattle themselves serve as attractors for the birds, and the creation of open, sparsely vegetated ground provides favorable foraging conditions for cowbirds and results in their increased use of riparian areas (Laymon, 1987).

<u>Nonriparian Aspen Woodland</u>

Aspen occurs in many types of western forests but only rarely occurs in extensive stands outside of riparian habitats, except in Colorado and Utah. Within conifer-dominated landscapes of the central and northern Rockies, aspen stands harbor more diverse breeding assemblages of birds than do coniferous habitats (Salt, 1957; Flack, 1976; Winternitz, 1976). Most studies of birds in nonriparian aspen stands have been conducted outside of the Northern Region (Debyle and Winokur,

1985), and indicate that species abundance and species richness vary greatly in conjunction with habitat structure, woody species composition, and landscape context (e.g., Flack, 1976).

In response to the recent commercial demand for aspen, a comprehensive examination of the effects of clear-cutting and consequent forest fragmentation for aspen avifaunas was initiated recently in the central Rockies (Finch and Reynolds, 1988). This well-designed long-term study includes wide geographic coverage by encompassing sites in Wyoming and Colorado, abundant replication, and three years of pretreatment analyses. Goals of the study are to evaluate the relationship of avian community composition and density to aspen stand age, woody species composition, and the size, shape, and landscape distribution of clearcuts.

Results of the first year of their study (Finch and Reynolds, 1988) comparing "large" (>40 ha) uncut aspen, mixed aspen/subalpine conifer (Engelmann spruce/subalpine fir), and uncut subalpine conifer stands indicated that nearly two thirds of the bird species that use aspen also occur in subalpine coniferous forest. As is typically the case for similar studies in coniferous forests, less than half of the species detected were sufficiently abundant to allow statistical comparisons among sites.

The average total number of birds per plot was highest in aspen, followed by mixed stands, followed by spruce/fir stands. Species richness followed the same pattern but did not differ statistically among the three stand types, although many species exhibited significantly different abundances among the three. In aspen stands, Red-naped Sapsucker, Western Wood-Pewee, Dusky Flycatcher, Tree Swallow, House Wren, Warbling Vireo, MacGillivray's Warbler, and Lincoln's Sparrow were significantly more abundant. Of these, the sapsucker, pewee, flycatcher, and wren were found almost exclusively in aspen stands. In addition, Black-capped Chickadee, Orange-crowned Warbler, and Brown-headed Cowbird were identified as potential obligate aspen associates by virtue of their abundance in aspen and mixed stands relative to spruce/fir stands. Only Red-breasted Nuthatch was more abundant in mixed stands than in either aspen or conifer stands. Olive-sided Flycatcher, Golden-crowned Kinglet, and Swainson's Thrush, were most abundant in the spruce/fir stands with the kinglet found nearly exclusively there (Finch and Reynolds, 1988).

Several species exhibited significant differences in relative abundance among aspen "subtypes"—stands with different understories (herbaceous, small shrub, and tall shrub). Thus understory characteristics within pure aspen stands can significantly affect species distributions among aspen stands.

Reduction of uncut aspen stands and conifer invasion of aspen stands were identified as likely to negatively affect the abundances of at least Red-naped Sapsucker and Warbling Vireo. Clear-cutting may improve some aspen subtypes by creating a mosaic of different structural stages, which will be differentially attractive across a suite of avian species. However, clear-cutting may degrade other aspen subtypes by removing or diminishing key habitat components required by obligate- or near-obligate aspen-nesting birds (Finch and Reynolds, 1988).

In a somewhat similar study in Colorado, Scott and Crouch (1988a) examined two stands in each of five overstory categories from pure aspen through mixed aspen/conifer to pure conifer. Although the study included two years of data, it is impossible to disentangle between-year differences from treatment effects because the data were combined. The authors noted summer livestock grazing on their plots but were unable to evaluate its possible impact; the potential influence of an obviously heterogeneous landscape also was not addressed.

Scott and Crouch (1988a) found that mixed stands had the highest avian species richness. Pure or nearly pure aspen harbored significantly greater abundances of House Wren, American Robin, Warbling Vireo, Yellow-rumped Warbler, and Dark-eyed

Junco. Significant negative correlations with pure or nearly pure aspen were found for Brown Creeper, Ruby-crowned Kinglet, Hermit Thrush, Western Tanager, and Pine Siskin—species that are normally associated with coniferous forests.

In a complementary study in Colorado, Scott and Crouch (1988b) examined three sizes of clearcuts (all under 7 ha) with 10 replicates each across five age-classes (6, 7, 8, 9, and 10 years postharvest), and included 30 plots in uncut aspen. Unfortunately, the study lasted only one season, and it is difficult to evaluate the landscape context and size of "uncut" controls. Not many differences in relative abundances were found among species that occupy a range of successional stages. Two species (Hermit Thrush and Solitary Vireo) normally associated with more mature habitats were significantly more abundant in the oldest clearcuts. However, birds that occupy large territories and many less common species were not sampled in numbers sufficient to determine their responses to clear-cutting.

The importance of considering the influence of surrounding coniferous forest on the composition of deciduous forest avifaunas is well illustrated by the work of Li and Martin (1991). In the southern Rockies of Arizona, they examined aspen with a coniferous overstory in a mixed-conifer forest dominated by ponderosa pine. Conifer snags were rare both in the ponderosa pine forest and in the aspen/pine woodland. As a result, aspen harbored an extremely diverse (14 species) assemblage of cavity-nesters (including Red-naped and Williamson's Sapsuckers, Cordilleran Fly-catcher, Brown Creeper, House Wren, and Western Bluebird), several of which normally are associated only with coniferous forest.

<u>Deciduous Nonriparian Draws</u>

Prairie thickets consisting of upland native woodlands (commonly known as "wooded draws") occur from eastern Montana through the Dakotas as islandlike habitats surrounded by prairie grasslands and agricultural fields. Wooded draws are characterized by small trees, predominantly green ash *(Fraxinus pennsylvanica)*, American elm *(Ulmus americana)*, and boxelder *(Acer negundo)* with substantial shrub understories. These woodlands are often subjected to intensive livestock grazing (Faanes, 1987) and are widely threatened by energy development activities, especially by strip-mining for coal (Hopkins, Cassel, and Bjugstad, 1986).

Although they cover less than 1 percent of the northern Great Plains (Bjugstad and Sorg, 1984), wooded draws harbor disproportionately rich avifaunas compared with other plains habitats. For example, in southeast Montana, Dubois (1979) found that deciduous draws supported the highest number of species and greatest breeding abundances of birds when compared with grasslands, savannah, and ponderosa pine forests in the region. Similarly, Hopkins, Cassel, and Bjugstad (1986) evaluated the breeding bird communities of four woodland types in western North Dakota and found the highest breeding densities in wooded draws; only riparian cottonwood habitats supported a larger number of breeding species.

No long-term studies of avian communities in wooded draws have been conducted. The only project that lasted for at least three years (Hopkins, Cassel, and Bjugstad, 1986) found substantial between-year differences, emphasizing the need for a longer-term perspective to gain understanding of population dynamics in these woodlands. Most of the work to date suffers from either insufficient replication (Hopkins, Cassel, and Bjugstad, 1986) or insufficient duration (Faanes, 1982, 1983, 1987).

Although only two sites in each of four woodland types were surveyed, Hopkins, Cassel, and Bjugstad (1986) found that Cooper's Hawk, Long-eared Owl, and Mountain Bluebird nested only in wooded draws and that many Neotropical migrants (Red-eyed Vireo, Black-and-white Warbler, Yellow-breasted Chat, American Redstart, Lazuli Bunting, Rufous-sided Towhee, Lark Sparrow, and American Goldfinch [also Black-capped Chickadee]) were significantly more abundant in ash woodlands than in juniper, pine, or even cottonwood habitats.

Only Faanes (1987) attempted to analyze wooded draw avifaunas in a landscape context by examining the dispersion pattern of draws relative to one another and by seeking area-related relationships for measures of avian community characteristics such as diversity and richness. No significant relationships were found between habitat area and either species diversity or richness (although species richness was found to increase with increasing area of ash woodlands in several theses cited by Hopkins, Cassel, and Bjugstad [1986]). Although Faanes (1987) provided no area data, it seems likely that the range of habitat sizes was simply too small to display any area effects across these inherently small woodlands.

The best assessment of livestock grazing impacts on wooded draw avifaunas was conducted over two years in northwestern South Dakota (Hodorff, Sieg, and Linder, 1988). The primary impact of cattle was the creation of open-canopy stands that consisted of a low shrub layer, a sparse overstory provided by decadent trees, an herbaceous layer of invasive, mostly alien species, and the complete absence of intermediate vegetation layers. In contrast, undisturbed stands were structurally complex and characterized by a closed canopy, varied size- and age-classes of trees and shrubs, and a significant herbaceous layer of native sedges, grasses, and forbs. Cattle grazing precluded woody plant recruitment by trees and tall shrubs in the open stands—a result that is strikingly similar to the effects of cattle grazing in riparian woodlands. The lack of successful reproduction and recruitment by trees and shrubs was leading to conversion of these woodlands into grass-forb communities.

Both breeding season and migration avifaunas were surveyed by Hodorff, Sieg, and Linder (1988). In a comparison between closed-canopy and open-canopy stands, overall numbers of birds were significantly greater in closed-canopy stands. Among Neotropical migrants, Orange-crowned Warbler, Rufous-sided Towhee, and American Goldfinch were significantly more abundant during the breeding season in closed-canopy stands (as were Great Horned Owl, Black-capped Chickadee, and Field Sparrow). Only the Western Meadowlark was significantly more abundant in open-canopy stands. During migration, *Empidonax* flycatchers, Swainson's Thrush, American Robin, Wilson's Warbler, and Dark-eyed Junco were significantly more abundant in the undisturbed stands (of these four species, only the robin was present during the breeding season).

Similarly in western North Dakota, Faanes (1987) documented a significant correlation between avian species diversity and foliage volume in the "high ground layer," which consisted of taller grasses and forbs, larger woody seedlings, and young shrubs. Faanes (1987) points out that this layer is the first to be impacted by cattle grazing and trampling in wooded draws.

Within the Northern Region, there appears to be significant geographic variation in the species composition of breeding bird communities in prairie thicket habitats. A long-term assessment of Neotropical migrant breeding populations across the full geographic spectrum of these woodlands is urgently needed in view of the potential loss of these habitats to increased levels of agricultural, industrial, and energy development (Hopkins, Cassel, and Bjugstad, 1986). The potentially great importance of these woodlands to migrating birds (Hodorff, Sieg, and Linder, 1988) adds further urgency to this need.

Shelterbelts

Shelterbelts are woodlands consisting of rows of trees and shrubs that form islands of woody vegetation surrounded by a sea of agricultural fields and native grasslands. Although superficially similar to wooded draws by virtue of their islandlike distribution and linear configuration, shelterbelts are planted, not natural, habitats. Shelterbelts occur as multirow plantings or as single-row windbreaks, and are widely distributed across the northern plains.

The extensive creation of shelterbelts across the Great Plains during this century, along with the spread of riparian woodlands that resulted from managed stream flows, provided dispersal corridors for forest birds. Many of these species were able to traverse the former ecological barrier presented by the grasslands of the Great Plains. The result was a mixing of species that were separated historically and produced an increased "cosmopolitism" of the region's avifauna (Knopf, 1986, 1992).

On a local scale in many parts of the Great Plains, shelterbelts are frequently the only significant woodlands. It is thus not surprising that birds often make considerable use of these habitats. Hopkins, Cassel, and Bjugstad (1986) reported greater total bird densities in shelterbelts of western North Dakota than in other types of woodlands, but lower species richness than found in other deciduous woodlands. Similarly, Faanes (1982) found higher total densities of breeding birds in shelterbelts than in any other habitat surveyed in east-central North Dakota (although Mourning Doves accounted for nearly 30 percent of the total).

Martin (1980) examined 69 shelterbelts over two years in South Dakota and found that total abundance and number of species were highly correlated with shelterbelt area during spring migration and during the breeding season. Several theses and unpublished reports cited by Hopkins, Cassel, and Bjugstad (1986) similarly reported that bird species richness and density increased with shelterbelt area and age during the breeding season in North Dakota. The importance of shelterbelts to breeding populations of declining or less common Neotropical migrant species is likely overstated by these general relationships, however. For example, Common Grackle, Mourning Dove, American Robin, and House Sparrow were four of the five most abundant species in the shelterbelts analyzed by Martin (1980). In western Minnesota, Yahner (1983) found only three species that were typical of the region's historic grasslands among the 47 most abundant species using shelterbelts.

Shelterbelts are viewed as potential substitutes for the loss of native upland woodlands in the northern plains (Hopkins, Cassel, and Bjugstad, 1986). While it is clear that birds make significant use of shelterbelts during the breeding season, no studies have examined the relative nesting success of these populations. Shelterbelts are consummate edge habitats embedded in essentially agricultural landscapes; as such, they are likely to attract primarily colonizing, edge-tolerant species of birds, and to provide easy nest access to avian and mammalian nest predators and the brood-parasitic cowbird. There is a distinct possibility that shelterbelts may function as population sinks (Robinson, 1992) or "ecological traps" (Gates and Gysel, 1978) for breeding Neotropical migrants. Before the adequacy of these habitats as substitutes for wooded draws can be evaluated, there must be detailed studies of avian productivity and demography carried out in shelterbelt habitats (and, in wooded draws, as well).

Grasslands and Shrubsteppe

Although native prairie grasslands and shrubsteppe support fewer species of birds than do forests, no other habitats in the Northern Region have as great a proportion of their species in decline (Paige, 1990). These widespread declines are linked directly to the tremendous reduction in acreage of these habitats caused by agricultural conversion to croplands and by degradation as a result of livestock overgrazing (Bock et al., in press).

Grasslands

Three types of native grasslands occur on the Great Plains: shortgrass, mixed-grass, and tallgrass prairie (Axelrod, 1985). Grasslands of the Northern Region consist primarily of mixed-grass prairie, with tallgrass prairie confined to the eastern Dakotas (Sims, 1988). Although total biomass is dominated by grasses and sedges, forbs and dwarf shrubs actually comprise a far greater number of the plant species that occur in these habitats. The total number of plant species found in grasslands increases with increasing length of growing season and total annual precipitation, and where topography is more varied and human-induced disturbance is minimal (Sims, 1988). Only a few remnants of tallgrass prairie remain uncultivated in the northern plains (Sims, 1988).

The mixed-grass prairie is largely a blend of species found in tallgrass and shortgrass prairies (Sims, 1988). This grassland occupies the western Dakotas, northeastern Wyoming, eastern Montana, and the southern portions of the central Canadian provinces. The mixture of intermediate-, short-, and occasional tallgrass species, a large number of forbs, and scattered low shrubs produces the richest plant diversity of all the grasslands found in the Great Plains (Barbour, Burk, and Pitts, 1980).

The species composition of mixed-grass prairies fluctuates more than that of other prairie habitats in response to extreme variation in total annual precipitation, fire suppression, assorted regimens of livestock grazing, and the generally greater topographic heterogeneity of areas occupied by mixed-grass prairie (Sims, 1988). With increasing drought and livestock grazing practices that further intensify aridity, the dominant plant species shift toward the more drought-tolerant species of shorter stature that are typical of shortgrass prairie. With increasing moisture and reduced grazing, plant composition shifts toward species more typical of tallgrass prairie. Although the dynamics of species fluctuations in the mixed-grass prairie are functions of climate, the magnitude of these changes is largely governed by grazing intensity (Sims, 1988).

Fire is an inherent and requisite component of natural cycles in these grasslands (Axelrod, 1985). Prior to the advent of modern agriculture and active fire suppression on the plains, fire frequency was greatest in tallgrass prairie (estimated at every 3–10 yrs), with longer intervals between fires in mixed-grass prairie and even longer intervals in shortgrass prairie (Daubenmire, 1968; Higgins, Kruse, and Piehl, 1986).

Drought, fire, and ungulate grazing have always been the primary ecological and evolutionary forces shaping the dynamics of prairie grasslands (Anderson, 1982; Hobbs and Huenneke, 1992). These forces combined in the past to create a large-scale mosaic of disturbance-generated habitats across the Great Plains. The size of habitat "patches" created by these disturbances (especially by fire) was likely to have been vastly greater than the size of modern disturbance patches because of the high degree of fragmentation that characterizes today's plains landscape. Given the highly fragmented nature of much of the remnant native prairie, it is surprising that so little attention has been paid to the biological implications of grassland fragmentation. Only two studies explicitly examine this question for prairie-nesting birds, one focused on songbirds (Johnson and Temple, 1986, 1990) and the other on waterfowl (Nelson and Duebbert, 1974).

Johnson and Temple (1986, 1990) found that rates of cowbird parasitism and nest predation were higher on nests of Clay-colored, Savannah, and Grasshopper Sparrows, Bobolink, and Western Meadowlark in small (16–32 ha) versus large (130–486 ha) fragments of undisturbed tallgrass prairie in Minnesota. Nest predation rates were lower for nests on large fragments, in areas more than 45 m from a wooded edge, and in vegetation that had been burned within the past three years. Rates of brood parasitism were lower on nests that were far from wooded edges. For all five species, the areas with the highest nest densities were not the areas with the highest nest productivity—a clear example of the potential for being misled by simply using breeding territory density as a measure of habitat quality or suitability for nesting populations of Neotropical migrants. Johnson and Temple suggest that management of tallgrass prairie to maximize nest productivity should provide large, regularly burned (at least every three years) prairies that are devoid of wooded edges such as shelterbelts or clumps of invading trees.

In an earlier study, Nelson and Duebbert (1974) similarly found higher nesting success for waterfowl on large (32–48 ha) blocks of upland vegetation compared to blocks of 16 ha or fewer, due to higher rates of nest predation in the smaller fragments. They too concluded that management of larger rather than smaller fragments of prairie habitat will provide for greater avian productivity.

Effects of Livestock Grazing

In the northern plains, the relative impact of grazing must be viewed in the context of the coevolution of these ecosystems with the grazing pressures exerted by vast herds of nomadic bison, elk, and pronghorn and huge colonies of black-tailed prairie dogs (Krueger, 1986; Whicker and Detling, 1988). The evolutionary history of grazing on the northern plains provides considerable insight toward understanding the relative ability of these ecosystems to support grazing by livestock (Mack and Thompson, 1982). The combination of grazing history and drought tolerance is a useful predictor of the grassland community's response to grazing pressures (Milchunas, Sala, and Lauenroth, 1988).

Grazing by domestic livestock differs fundamentally from grazing by bison or other native ungulates because herds of livestock remain in a restricted area rather than shift across vast landscapes. Domestic livestock exert prolonged and intense grazing pressure that varies with the number of animals per unit area and with the season and duration of their presence. One effect of livestock on northern plains ecosystems has been to greatly reduce fire frequency and intensity by consuming the fine fuels needed to transport fire. The result has been increased invasion by woody plants and the general replacement of intermediate- and tallgrass species by more grazing-tolerant species of shorter stature. In tallgrass prairie, where fires play a critical role in excluding woody vegetation (Gibson and Hulbert, 1987), livestock grazing has converted many areas into woodlands or shrublands. In mixed-grass prairie, fire suppression and cattle grazing result in increased abundance and size of two common shrubs, wolf- or snowberry *(Symphoricarpos occidentalis)* and silverberry *(Elaeagnus commutata)* (Arnold and Higgins, 1986).

The responses of grassland-nesting birds to the effects of livestock grazing in plains grasslands are reviewed in detail by Bock et al. (in press). Although several studies have attempted to compare breeding bird communities among different grazing regimens, virtually all comparisons are across qualitative assessments of grazing intensities (i.e., heavy, moderate, and light). Moreover, there are no long-term studies for any grassland-nesting species and no studies designed explicitly to compare population productivity among undisturbed, grazed, burned, and hayed grasslands. It is therefore impossible to evaluate results meaningfully across studies. In spite of this significant limitation, some consistencies do emerge for several species in the Northern Region. The conclusions reached by Bock et al. (in press) do not differ markedly from those presented by Kirsch, Duebbert, and Kruse (1978) in their earlier review: livestock grazing (and mowing of hay crops) adversely impacts most (but not all) species of upland-nesting birds.

Perhaps the most important generalization that can be made about the responses of nesting Neotropical migrants to livestock grazing in the region's grasslands is that no single generalization can be made for these species as a group. All grassland types support some species that are grazing tolerant or even grazing dependent, as well as some species that are moderately to extremely grazing intolerant—which is consistent with the mosaiclike distribution of disturbance patches that these grasslands presented to their evolving avifauna. Some species vary in their responses depending on the type of grassland involved (short, mixed, or tall; e.g., see Kantrud, 1981) and depending on the intensity and season of grazing. Other species do not conform to this pattern and instead respond rather uniformly. For example, Horned Larks are invariably favored by the creation of open areas through grazing. At the opposite end of the response spectrum, Northern Harrier and

Short-eared Owl nest in dense vegetation found only in ungrazed areas (Duebbert and Lokemoen, 1977). The latter two species frequently nest in association with stands of shrubs (especially snowberry) in undisturbed native grasslands (Kantrud and Higgins, 1992).

Most upland shorebirds generally prefer relatively short and sparse grasslands (Kantrud and Higgins, 1992), ranging from minimal vegetation for Mountain Plover to well-vegetated sites for Upland Sandpiper. Among these shorebirds, we have meaningful data on nest success (rather than simply on nest occurrence) only for Upland Sandpipers, which experience significantly greater nesting success in undisturbed and in burned mixed-grass prairie compared with grazed habitats (Kirsch and Higgins, 1976; Kantrud and Higgins, 1992; but see Bowen and Kruse, 1993). Both Marbled Godwit and Willet appear to nest preferentially in native prairie rather than in grasslands seeded with exotic species (Kantrud and Higgins, 1992). The presence of livestock appears to deter nesting by most shorebirds (Kantrud and Higgins, 1992; Bowen and Kruse, 1993).

Songbirds similarly exhibit an array of species-specific responses: Savannah Sparrow, Baird's Sparrow, and Bobolink (as well as the declining, non-Neotropical, Sharp-tailed and Le Conte's Sparrows) almost invariably respond negatively to any level of livestock grazing in mixed-grass prairie (e.g., Maher, 1979). Disturbance either by mowing or cattle grazing is known to reduce or eliminate populations of Sprague's Pipit and Baird's Sparrow in southern Alberta (Owens and Myres, 1973). At the other extreme, Horned Lark, Lark Sparrow, and McCown's Longspur generally respond positively to moderate and sometimes to heavy grazing pressure (e.g., Owens and Myres, 1973; Maher, 1979).

Another group of species responds positively or negatively to livestock grazing depending on the type and condition of grassland. Species that appear to favor intermediate stature and cover respond positively to grazing in tallgrass prairie and more robust mixed-grass prairie, but respond negatively in sparser mixed-grass habitats (e.g., Sprague's Pipit, Grasshopper Sparrow, and Chestnut-collared Longspur) (Bock et al., in press) and in shortgrass prairie (Ryder, 1980).

Impacts of Pest Management

Since 1977, the USDA Animal Plant Health Inspection Service (APHIS) has conducted an extensive grasshopper control program across many portions of the western United States on public and private rangelands and croplands. This program includes extensive aerial application of broad spectrum insecticides across tens of thousands to hundreds of thousands of hectares annually in the Northern Region. These control programs indiscriminately reduce populations of a wide range of insects and other invertebrate species. The effects (both direct and indirect) on grassland birds are far from clear, and several studies are now in progress to evaluate some of the potential problems for selected species (McEwen, George, and Petersen, 1990). The wholesale reduction of prey populations for insectivorous grassland-nesting birds across large areas should be of great concern to land managers.

Summary and Recommendations

The continued loss of habitat poses the greatest threat to breeding populations of Neotropical migrants in native prairie grasslands. Relatively little undisturbed mixed-grass prairie and very little native tallgrass prairie of any sort remain. As recommended by Bock et al. (in press), there is an urgent need to dramatically increase the amount of public rangeland from which all livestock are permanently excluded. There is no shortage of grazed and hayed lands for those species that benefit from these activities. By comparison, habitat for species with breeding requirements that are not compatible with grazing and haying is exceedingly rare and continues to diminish.

In the northern plains, the native prairie on the National Grasslands offers the most viable opportunity for preservation of breeding populations of grazing-intolerant Neotropical migrants. These areas could be managed by a return to natural ecosystem functioning through the use of prescribed burning in conjunction with minimal livestock grazing. In contrast to most current practices, livestock use of allotments would require separation by long (25–50 year) rotations to allow reasonable recovery of these areas (Bock et al., in press).

Chronic livestock grazing on shortgrass and mixed-grass prairie alters the structure and composition of plant communities in ways that greatly enhance conditions for population outbreaks of grasshoppers (see e.g., Quinn and Walgenbach, 1990, and references therein). Benefits of the proposed management scenario for land managers concerned with maintaining or enhancing populations of Neotropical migrants in native grasslands include the elimination of pesticide applications under the APHIS program. As rangeland conditions improve, these grasslands will become increasingly unlikely to support grasshopper outbreaks, and will provide conditions that should allow predation by grassland birds (and other natural enemies) to effectively maintain grasshopper populations at low or moderate densities (Fowler et al., 1991).

An additional avenue for maintaining and restoring grassland-nesting Neotropical migrants is offered by the federally sponsored Conservation Reserve Program (CRP). The CRP was designed to encourage landowners to convert formerly tilled cropland into grassland. Millions of hectares of CRP grasslands have been planted on the Great Plains since 1985, although most of these lands are planted with exotic grasses (Joyce et al., 1991). Even so, CRP grasslands are of far greater use to native wildlife than the croplands that they replace (Kantrud and Higgins, 1992). By modifying the current program to require plantings exclusively of native vegetation and encouraging their long-term maintenance, the amount of suitable breeding habitat for populations of grassland-inhabiting Neotropical migrants (and many other wildlife species, as well) could be increased dramatically.

Shrubsteppe

Sagebrush-dominated (*Artemisia* sp.) shrubsteppe is distributed over much of the Intermountain West—the area bounded by the mountains of the Sierra and Cascades on the west and the Rocky Mountains on the east. Although the avifauna of shrubsteppe is one of the most species-poor in the United States, it is relatively distinctive (Wiens, 1989a). Three species of Neotropical migrants (Sage Thrasher, Sage Sparrow, and Brewer's Sparrow) are almost exclusively associated with sagebrush-shrubsteppe, and two others (Green-tailed Towhee and Vesper Sparrow) are largely associated with this habitat (Baker et al., 1976).

Unlike native prairie grasslands, shrubsteppe plant communities did not coevolve with nomadic herds of bison or other large ungulates (Mack and Thompson, 1982). The introduction of cattle and sheep to areas of shrubsteppe profoundly altered the vegetation of these habitats and resulted in a virtually complete loss of the perennial grasses that grew among the scattered shrubs. These landscapes have been converted by livestock grazing (and fire suppression) into relatively much denser shrub communities, with the interstitial native bunchgrasses replaced by bare ground and exotic grasses and forbs.

The combination of intense livestock grazing and active fire suppression facilitates the spread of shrubs into grasslands and increases the density of shrub coverage. As a result, shrubsteppe habitat in the Northern Region, if defined by its structure rather than by shrub species composition, now occurs from portions of the western plains in the Dakotas through much of eastern and central Montana and across northern Wyoming.

Extensive studies by Wiens, Rotenberry, and their colleagues document the dynamic nature of shrubsteppe avifaunas (Wiens and Rotenberry, 1981; Wiens, Rotenberry, and Van Horne, 1986; Rotenberry and Wiens, 1989; Wiens, 1989a) and demonstrate the high degree of unpredictability in the fluctuations of shrubsteppe bird populations. Against this template of inherent change, it is unfortunate that no long-term, well-replicated studies have been conducted to compare avifaunas in grazed and ungrazed shrubsteppe communities. The extreme scarcity of ungrazed shrubsteppe that is not dominated by exotic vegetation makes it highly unlikely that we will ever be able to assess the historic effects of livestock grazing on these avifaunas. Based on the changes that grazing produces in shrubsteppe vegetation and given the nesting requirements for many of the birds associated with these habitats, Bock et al. (in press) attempted to provide a reasoned set of predictions for responses by some shrubsteppe-typical species to grazing. Nearly all of the relevant studies that have been conducted occurred outside the Northern Region.

Responses by some species will vary as the grazing-altered plant community changes through time. For example, Burrowing Owl and Horned Lark initially will respond positively to decreased herbaceous cover and increased bare ground produced by livestock grazing, but eventually will respond negatively to the increasing shrub coverage that grazing brings about. Increased bare ground for foraging and increased shrub coverage for nesting and perching should positively favor both Sage Sparrow and Brown-headed Cowbird. Aside from studies of nesting Sage Thrashers (Rich and Rothstein, 1985), little is known about the interactions of cattle, cowbirds, and nest parasitism in shrubsteppe habitats.

There is virtually no information concerning the responses of raptors to grazing in shrubsteppe. Bock et al. (in press) suggest that Golden Eagles may respond positively to grazing by virtue of the association between their primary prey (jackrabbits) and shrubby habitats. Similarly, the need for dense ground cover for nest sites and the association between rodent prey and herbaceous ground cover likely produce negative responses to grazing by Northern Harrier, Ferruginous Hawk, and Short-eared Owl. The requirement of grass and forb cover for nesting and foraging likely translates into negative responses to grazing for Long-billed Curlew, Vesper, Savannah, and Grasshopper Sparrows, and Western Meadowlark.

As indicated by Bock et al. (in press), the great majority of shrubsteppe-nesting bird species are likely to be impacted negatively by livestock grazing. The extreme modification of vegetation structure and species composition in shrubsteppe as a result of livestock grazing creates communities that are depauperate both in plant and avian species.

Considerable effort has been expended to "control" sagebrush in order to provide increased forage for cattle in overgrazed shrubsteppe where most of the native grass component has been lost. Herbicidal treatment, mechanical alteration (plowing, chaining, and disking), and burning of extensive, decadent stands have all been used toward this end.

The effect of herbicidal spraying on shrubsteppe avifaunas appears to depend upon the relative proportion of sagebrush that is killed and the extent to which the dead shrubs remain standing. In Montana shrubsteppe, herbicidal spraying that killed 50 percent or less of the sagebrush produced no differences in nesting densities of either Brewer's or Vesper Sparrows (Best, 1972), which together comprised between 75 and 90 percent of the breeding birds (Feist, 1968). In plots in Montana and Wyoming where 100 percent of the sagebrush was killed, the shrub-nesting Brewer's Sparrow declined by more than 50 percent in the first year following spraying (Best, 1972; Schroeder and Sturges, 1975), while the ground-nesting Vesper Sparrow was unaffected. Subsequent studies in Montana found that Brewer's Sparrows had disappeared almost completely by the fifth year following spraying (described by Baker et al., 1976).

Although fire apparently occurred less frequently in shrubsteppe habitats than in plains grasslands, it nevertheless was an integral part of the normal functioning of shrubsteppe ecosystems. Prior to the appearance of domestic livestock, fire presumably created a landscape-scale mosaic of habitats ranging from relatively bunchgrass-dominated communities in recently burned patches to relatively shrub-dominated communities in areas that had not burned for several decades.

In today's much-altered landscape, the inability of sagebrush to resprout after being burned has led to the replacement of shrubsteppe in many places by European annual grass communities in the wake of repeated fires (West, 1988). Complete removal of the standing shrub community by fire or by other means, followed by complete replacement with grasses, results in a species-poor, sparsely-nesting "community" of breeding birds composed almost entirely of Horned Larks or Western Meadowlarks (Reynolds and Trost, 1981; Bock and Bock, 1987; T. D. Rich, pers. comm.). In contrast, where fire produces a spatially heterogeneous mosaic of burned and unburned patches, the breeding avifauna appears to be little affected (Petersen and Best, 1987). Land managers seeking to return dense, decadent stands of shrubsteppe to more natural conditions should strive to produce landscape-scale mosaics of burned and unburned patches if this can be achieved without risking wholesale invasion of cheatgrass *(Bromus tectorum)* and other exotic herbaceous species (Hobbs and Huenneke, 1992).

HABITAT PROFILES

The following habitat profiles are schematic representations intended to illustrate the impacts of selected management activities on Neotropical migrant landbirds in the northern Rockies and Great Plains. The presence and absence of species depicted in the profiles illustrate the changes in avian community composition that occur under different management actions, and are based on research described in the preceding section covering management activities and impacts. Names of species appear in the drawings to illustrate nesting and/or foraging locations used by the birds in these habitats. Species designated in parentheses are not Neotropical migrants but are key species that are significantly associated with the habitat.

Figure 4. Old-growth coniferous forest characterized by abundant live trees of varied age-classes, a relatively closed canopy, many standing snags of very large diameter, large-diameter downed and dead woody debris, and a scattered but dense growth of herbaceous plants and shrub thickets. The indicated species represent a composite of significant associates of old-growth Douglas fir and ponderosa pine forests in the region.

Thinned, rotation-age coniferous forest characterized by even-age trees spaced more widely than in the old growth, a relatively open canopy, sparse scattered grasses, forbs, and shrubs, and the complete absence of large-diameter downed and dead woody debris. Standing snags are rare and large standing snags are entirely absent because large live trees were not retained at harvest, thus precluding snag recruitment as the stand aged. The structural simplification of stands managed in this way results in the absence of nearly all birds normally associated with old growth.

Clear-cut coniferous forest managed to maximize structural diversity by retention of several large live trees to provide nesting and foraging substrates and for snag recruitment in the future, retention of standing snags of varied sizes, retention of all deciduous trees and shrub thickets, and retention of slash piles to provide nesting and foraging habitat. Retaining these features more closely simulates the conditions following a natural disturbance (such as fire) and facilitates the pace of successional recovery.

Figure 4.

OLD-GROWTH CONIFEROUS FOREST

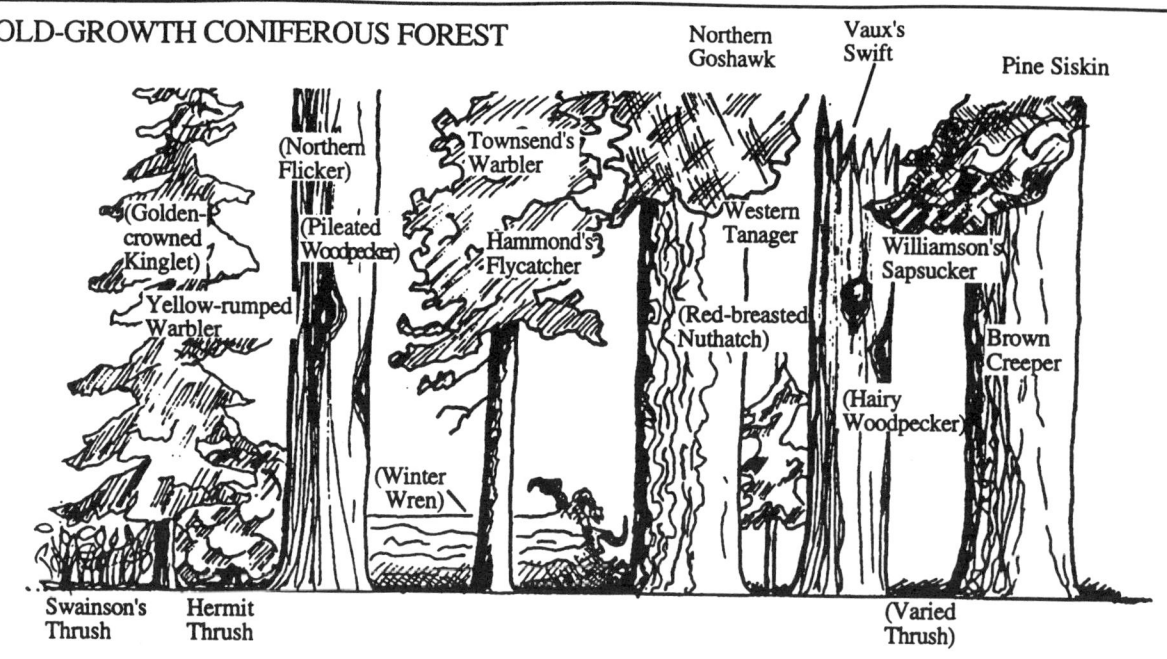

THINNED, ROTATION-AGE CONIFEROUS FOREST

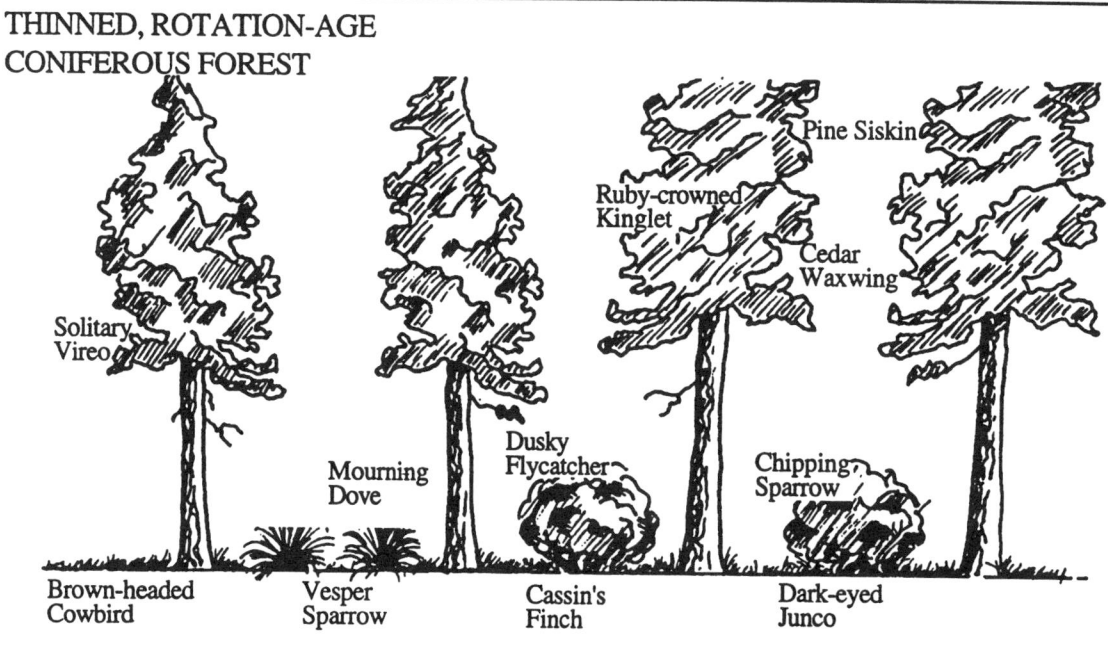

EARLY SUCCESSION CLEAR-CUT CONIFEROUS FOREST

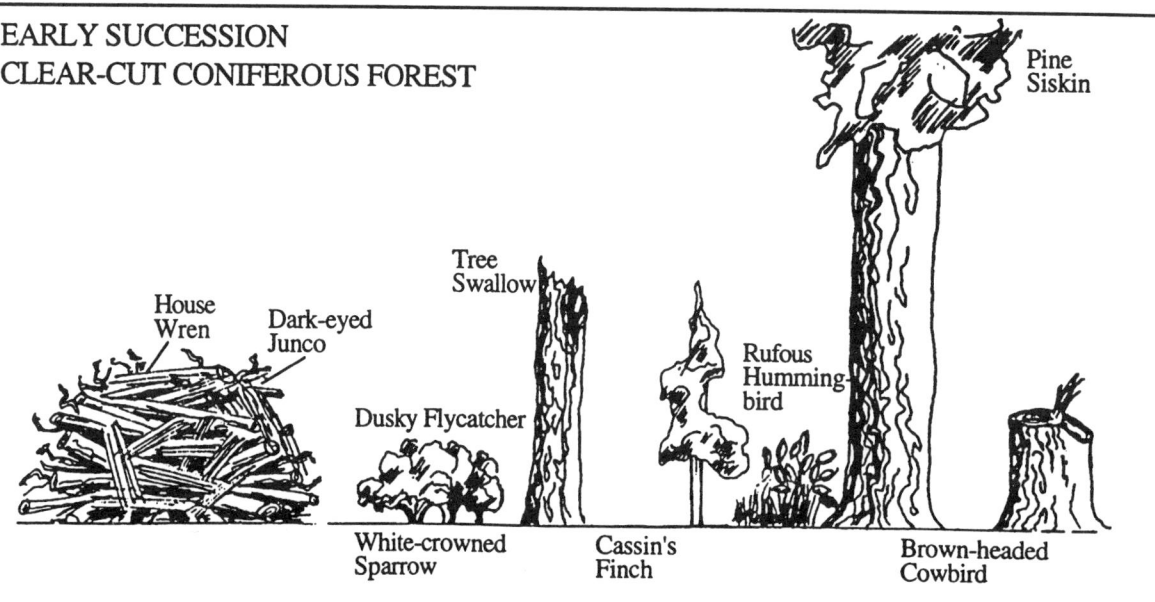

Figure 5.

RIPARIAN WOODLAND, UNGRAZED

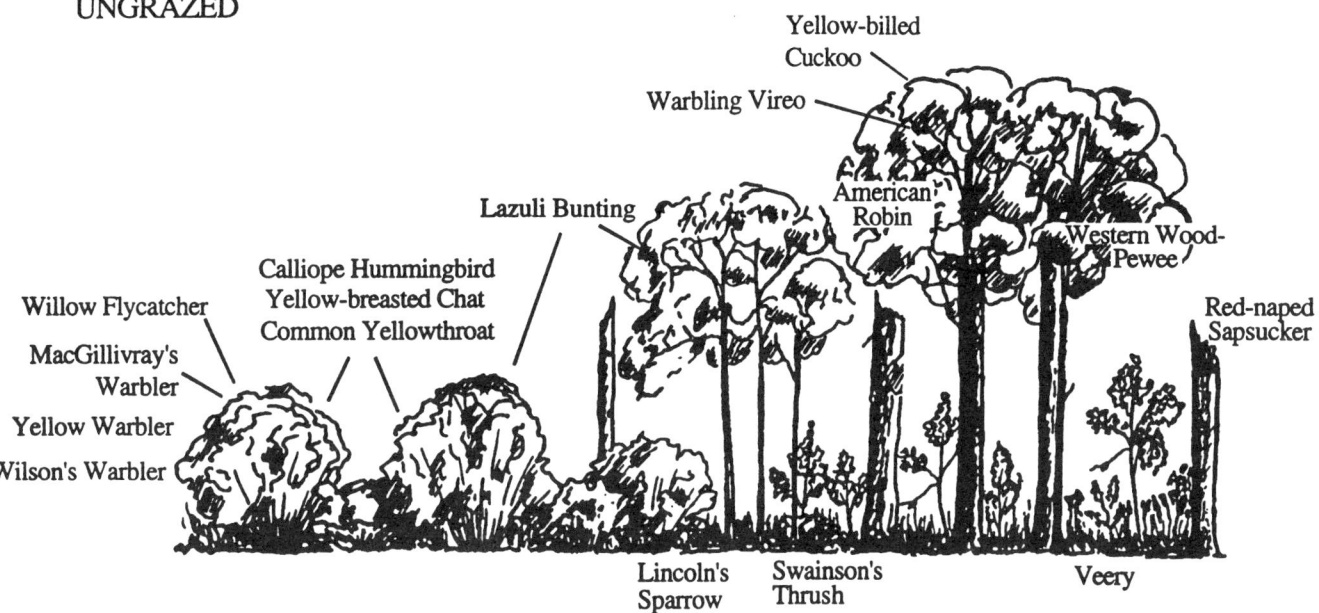

Riparian woodland without livestock grazing, characterized by robust trees and shrubs (from left to right, willow, aspen, and cottonwood) of varied sizes and age-classes, abundant standing snags, and a dense herbaceous layer of forbs, sedges, and grasses. Tree overstories are relatively dense, and shrub thickets are dense and continuous. Recruitment of young trees and shrubs is evident.

RIPARIAN WOODLAND, HEAVILY GRAZED

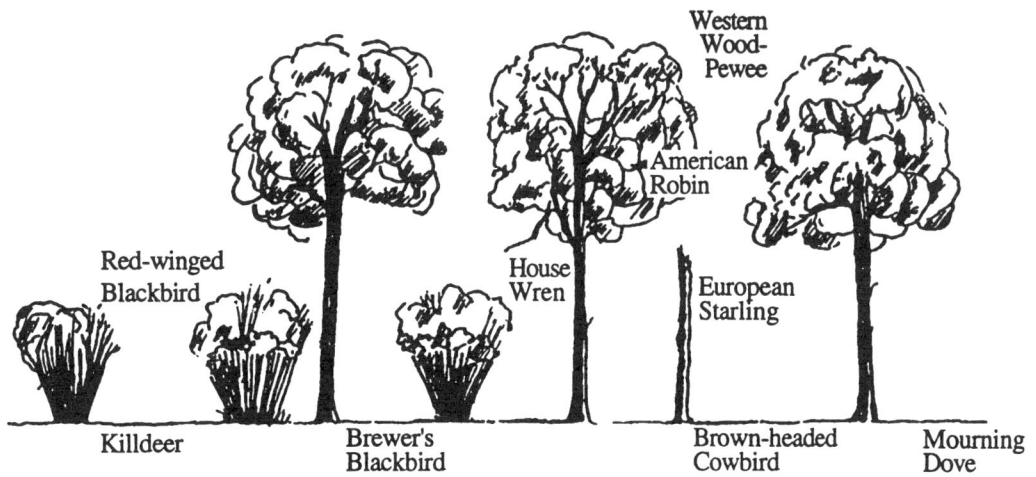

Heavily grazed riparian woodland, characterized by a sparse, decadent overstory of trees, scattered clumps of decadent, pedestaled shrubs, the replacement of most of the herbaceous layer by bare ground, and the complete absence of recruitment by woody species. Snags do not remain standing for long and are relatively scarce. Bird species composition in this habitat differs dramatically from the avifauna in the ungrazed, healthy riparian woodland. Most of the birds in the heavily grazed riparian are generalist species that occur widely in human-disturbed landscapes.

Figure 6.

MIXED-GRASS PRAIRIE/SHRUBSTEPPE,
UNGRAZED
PRAIRIE BURNED AT 5–10 YEAR INTERVALS

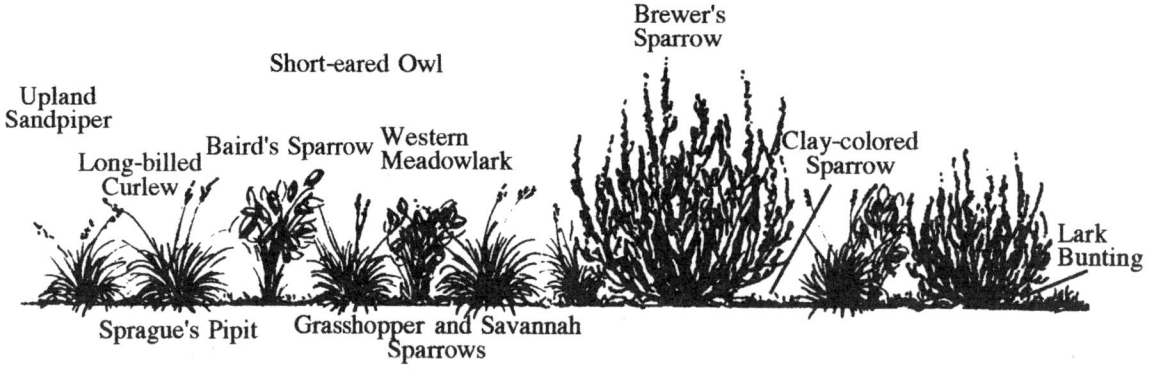

Ungrazed, regularly burned mixed-grass prairie, characterized by robust native bunchgrasses and forbs with relatively little bare ground; ungrazed shrubsteppe, characterized by shrubs of varying heights and native bunchgrasses growing between shrubs.

MIXED-GRASS PRAIRIE/SHRUBSTEPPE

HEAVILY GRAZED | MODERATELY GRAZED

Moderately grazed mixed-grass prairie, characterized by bunchgrasses of shorter stature and relatively more bare ground than occurs in ungrazed prairie; moderately grazed shrubsteppe, characterized by shrubs of more uniform size and greater density, and by loss of most interstitial bunchgrasses. As conditions change under moderate grazing, most bird species that are typical of ungrazed habitat are lost and a few new species are added.

Heavily grazed mixed-grass prairie, characterized by widely spaced bunchgrasses of short stature and much bare ground; heavily grazed shrubsteppe, characterized by decadent stands of shrubs that are uniform in size and are separated by areas of bare ground due to loss of all interstitial bunchgrasses. As conditions change from moderate to heavy grazing, bird species that are typical of ungrazed and moderately grazed conditions are lost and replaced mostly by generalist species that thrive in highly disturbed habitats.

EXPLANATION OF SPECIES ACCOUNTS

The following section contains individual accounts for the 144 species of Neotropical migrant landbirds that breed in the Northern Region. The information contained in each account is drawn from a wide range of source materials and was critically evaluated for applicability to populations occurring in the region. The information is not intended to be all-encompassing for species that have breeding populations outside of the region as well. For many birds, specific information pertaining to populations in the northern Rockies and Great Plains simply is not available, and I have tried to winnow through more general information to select only what may be applicable for the region's populations. I refer the reader to *The Birder's Handbook: A Field Guide to the Natural History of North American Birds* (Ehrlich, Dobkin, and Wheye, 1988) for broader perspectives and additional details of North American species.

These accounts are meant to give the reader a brief synopsis of the most pertinent information relevant to the conservation and management needs of each species. They are road maps of our knowledge about populations of Neotropical migrants in the Northern Region. These accounts bear an implicit invitation to gather the data needed to fill the all-too-obvious gaps in our "maps."

Preceding the species accounts are two tables that list all the species and their familial/subfamilial relationships, and indicate where each species winters. Table 1 lists the species in standard taxonomic order, grouped by family or subfamily, and provides a numeric code for wintering area, as follows:

1. Winters exclusively in South America
2. Winters exclusively in Mexico
3. Winters from Mexico southward
4. Winters from U.S. southward
5. Winters entirely within U.S. and Mexico
6. Some populations are Resident in the U.S. and some winter south of the U.S.

Likewise, Table 2 lists the species phylogenetically but is organized by wintering area.

Each species account consists of the following sections:

1. Seasonal residency status for populations within the region.

2. Wintering area code (from Tables 1 and 2).

3. The range of habitats utilized by nesting populations within the region, including specific nest location. Relative importance of different habitats to breeding populations is provided where such information is available.

4. Principal food habits and foraging techniques.

5. Status and management, including analy-sis of regional, western, and continent-wide population trends based on the USFWS Breeding Bird Survey database; information on sources of population problems; conservation status; ecological or behavioral peculiarities that may affect management; specific management impacts (both negative and positive), including potential for population enhancement; frequency of cowbird parasitism. Citations pertaining to management activities for the species are included within this paragraph.

Description of population trends within the Northern Region is presented qualitatively as the BBS data are frequently insufficient for meaningful statistical analysis. I use the term "significant" in the context of these population trend descriptions only where the described trend is statistically significant.

6. Further reading relating to ecological and behavioral information for the species. Only the most recent citations are given to provide the user with entry to the literature without producing an overwhelmingly large bibliography. The reader can then use the bibliographies of the cited references to work back into the older literature. Citations provided in the Status and Management paragraph are not repeated here.

7. At the bottom of each account is a detailed range map compiled from the most recent statewide atlases and distributional studies for the states covered. The maps illustrate the breeding season distribution of each species within the region. Where information could not be located or appeared contradictory, I inserted question marks to highlight the need for more accurate data. Although these maps are the most detailed to date for the region as a whole, they are still fairly coarse and certainly subject to revision with the acquisition of more extensive and more recent information.

Idaho distributions are based on recently compiled latilong maps (Stephens and Sturts, 1991). Montana distributions were mapped on a finer scale, by occurrences in quarter divisions within each latilong (Bergeron et al., 1992). Gaps in species' ranges in these two states appear on the range maps with straight edges as a result of close adherence to the latilong distributions. While a single empty block surrounded by shaded areas may not make immediate intelligible sense biologically, it nevertheless signifies that no verified records exist for that area.

North Dakota distributions are based on Stewart (1975), and supplemented where possible with recent publications dealing with specific areas of the state. South Dakota maps are based partly on the revised *Birds of South Dakota* (South Dakota Ornithologists' Union, 1991), and partly on Johnsgard (1979). The revised western range maps of Peterson (1990) and of the National Geographic Society (1990) were used in conjunction with a wide array of literature to fill in ranges for the remainder of the region.

The reader should bear in mind that most species are associated with specific habitats that may be relatively discontinuous or even quite rare across the Northern Region. In other words, most species are not distributed uniformly throughout the shaded portions of the range maps. In all cases, I attempted to evaluate the details of each species' distribution and conservatively excluded most "outlier" occurrences and records of isolated sightings that were not of recent origin.

Table 1. Neotropical Migrants in Taxonomic Order Including Wintering Area

Family/Subfamily	Species	Wintering Area
Cathartidae	Turkey Vulture	4
Accipitridae	Osprey	4
Accipitridae	Northern Harrier	4
Accipitridae	Sharp-shinned Hawk	4
Accipitridae	Cooper's Hawk	4
Accipitridae	Northern Goshawk	5
Accipitridae	Broad-winged Hawk	3
Accipitridae	Swainson's Hawk	1
Accipitridae	Red-tailed Hawk	4
Accipitridae	Ferruginous Hawk	5
Accipitridae	Golden Eagle	6
Falconidae	American Kestrel	4
Falconidae	Merlin	4
Falconidae	Peregrine Falcon	4
Falconidae	Prairie Falcon	5
Charadriidae	Mountain Plover	5
Charadriidae	Killdeer	4
Scolopacidae	Willet	4
Scolopacidae	Spotted Sandpiper	4
Scolopacidae	Upland Sandpiper	1
Scolopacidae	Long-billed Curlew	5
Scolopacidae	Marbled Godwit	4
Columbidae	Mourning Dove	4
Cuculidae	Yellow-billed Cuckoo	1
Cuculidae	Black-billed Cuckoo	1
Strigidae	Flammulated Owl	3
Strigidae	Burrowing Owl	4
Strigidae	Long-eared Owl	5
Strigidae	Short-eared Owl	5
Caprimulgidae	Common Nighthawk	1
Caprimulgidae	Common Poorwill	5
Apodidae	Chimney Swift	1
Apodidae	Vaux's Swift	3
Apodidae	Black Swift	3
Apodidae	White-throated Swift	4
Trochilidae	Ruby-throated Hummingbird	4
Trochilidae	Black-chinned Hummingbird	2
Trochilidae	Calliope Hummingbird	2
Trochilidae	Broad-tailed Hummingbird	3
Trochilidae	Rufous Hummingbird	2
Alcedinidae	Belted Kingfisher	4
Picidae	Lewis' Woodpecker	5
Picidae	Yellow-bellied Sapsucker	4
Picidae	Red-naped Sapsucker	4
Picidae	Williamson's Sapsucker	5
Tyrannidae	Olive-sided Flycatcher	1
Tyrannidae	Western Wood-Pewee	1
Tyrannidae	Eastern Wood-Pewee	1
Tyrannidae	Least Flycatcher	3
Tyrannidae	Hammond's Flycatcher	4
Tyrannidae	Dusky Flycatcher	5
Tyrannidae	Willow Flycatcher	3
Tyrannidae	Cordilleran Flycatcher	2
Tyrannidae	Eastern Phoebe	5
Tyrannidae	Say's Phoebe	5
Tyrannidae	Great Crested Flycatcher	3
Tyrannidae	Cassin's Kingbird	3
Tyrannidae	Western Kingbird	3
Tyrannidae	Eastern Kingbird	1
Alaudidae	Horned Lark	6
Hirundinidae	Purple Martin	1
Hirundinidae	Tree Swallow	4
Hirundinidae	Violet-green Swallow	4
Hirundinidae	Northern Rough-winged Swallow	4
Hirundinidae	Bank Swallow	1
Hirundinidae	Cliff Swallow	1
Hirundinidae	Barn Swallow	1
Certhiidae	Brown Creeper	6
Troglodytidae	Rock Wren	5
Troglodytidae	House Wren	5
Troglodytidae	Sedge Wren	5
Troglodytidae	Marsh Wren	5
Sylviinae	Ruby-crowned Kinglet	4
Muscicapinae	Eastern Bluebird	6
Muscicapinae	Western Bluebird	6
Muscicapinae	Mountain Bluebird	5
Muscicapinae	Townsend's Solitaire	5

Table 1. (continued)

Family/Subfamily	Species	Wintering Area	Family/Subfamily	Species	Wintering Area
Muscicapinae	Veery	1	Emberizinae	Chipping Sparrow	5
Muscicapinae	Swainson's Thrush	3	Emberizinae	Clay-colored Sparrow	5
Muscicapinae	Hermit Thrush	4	Emberizinae	Brewer's Sparrow	5
Muscicapinae	American Robin	6	Emberizinae	Vesper Sparrow	5
Mimidae	Gray Catbird	4	Emberizinae	Lark Sparrow	5
Mimidae	Northern Mockingbird	6	Emberizinae	Sage Sparrow	5
Mimidae	Sage Thrasher	5	Emberizinae	Savannah Sparrow	4
Motacillidae	American Pipit	4	Emberizinae	Lark Bunting	5
Motacillidae	Sprague's Pipit	5	Emberizinae	Baird's Sparrow	5
Bombycillidae	Cedar Waxwing	4	Emberizinae	Grasshopper Sparrow	4
Laniidae	Loggerhead Shrike	5	Emberizinae	Fox Sparrow	5
Vireonidae	Bell's Vireo	3	Emberizinae	Song Sparrow	6
Vireonidae	Solitary Vireo	4	Emberizinae	Lincoln's Sparrow	4
Vireonidae	Yellow-throated Vireo	4	Emberizinae	Swamp Sparrow	5
Vireonidae	Red-eyed Vireo	1	Emberizinae	White-crowned Sparrow	5
Vireonidae	Warbling Vireo	3	Emberizinae	Dark-eyed Junco	6
Parulinae	Tennessee Warbler	3	Emberizinae	McCown's Longspur	5
Parulinae	Orange-crowned Warbler	4	Emberizinae	Chestnut-collared Longspur	5
Parulinae	Nashville Warbler	4	Icterinae	Bobolink	1
Parulinae	Yellow Warbler	4	Icterinae	Red-winged Blackbird	6
Parulinae	Yellow-rumped Warbler	4	Icterinae	Western Meadowlark	6
Parulinae	Townsend's Warbler	4	Icterinae	Yellow-headed Blackbird	5
Parulinae	Black-and-white Warbler	4	Icterinae	Brewer's Blackbird	6
Parulinae	American Redstart	3	Icterinae	Orchard Oriole	3
Parulinae	Ovenbird	4	Icterinae	Northern Oriole	3
Parulinae	Northern Waterthrush	4	Fringillidae	Cassin's Finch	5
Parulinae	MacGillivray's Warbler	3	Fringillidae	Pine Siskin	6
Parulinae	Common Yellowthroat	4	Fringillidae	American Goldfinch	5
Parulinae	Wilson's Warbler	4			
Parulinae	Yellow-breasted Chat	4			
Thraupinae	Scarlet Tanager	1			
Thraupinae	Western Tanager	3			
Cardinalinae	Lazuli Bunting	5			
Cardinalinae	Indigo Bunting	3			
Cardinalinae	Dickcissel	3			
Cardinalinae	Rose-breasted Grosbeak	3			
Cardinalinae	Black-headed Grosbeak	5			
Emberizinae	Green-tailed Towhee	5			
Emberizinae	Rufous-sided Towhee	6			

Table 2. Neotropical Migrants in Taxonomic Order Organized by Wintering Area

Family/Subfamily	Species	Wintering Area
WINTERS IN SOUTH AMERICA		
Accipitridae	Swainson's Hawk	1
Scolopacidae	Upland Sandpiper	1
Cuculidae	Yellow-billed Cuckoo	1
Cuculidae	Black-billed Cuckoo	1
Caprimulgidae	Common Nighthawk	1
Apodidae	Chimney Swift	1
Tyrannidae	Olive-sided Flycatcher	1
Tyrannidae	Western Wood-Pewee	1
Tyrannidae	Eastern Wood-Pewee	1
Tyrannidae	Eastern Kingbird	1
Hirundinidae	Purple Martin	1
Hirundinidae	Bank Swallow	1
Hirundinidae	Cliff Swallow	1
Hirundinidae	Barn Swallow	1
Muscicapinae	Veery	1
Vireonidae	Red-eyed Vireo	1
Thraupinae	Scarlet Tanager	1
Icterinae	Bobolink	1
WINTERS IN MEXICO		
Trochilidae	Black-chinned Hummingbird	2
Trochilidae	Calliope Hummingbird	2
Trochilidae	Rufous Hummingbird	2
Tyrannidae	Cordilleran Flycatcher	2
WINTERS FROM MEXICO SOUTHWARD		
Accipitridae	Broad-winged Hawk	3
Strigidae	Flammulated Owl	3
Apodidae	Vaux's Swift	3
Apodidae	Black Swift	3
Trochilidae	Broad-tailed Hummingbird	3
Tyrannidae	Least Flycatcher	3
Tyrannidae	Willow Flycatcher	3
Tyrannidae	Great Crested Flycatcher	3
Tyrannidae	Cassin's Kingbird	3
Tyrannidae	Western Kingbird	3
Muscicapinae	Swainson's Thrush	3
Vireonidae	Bell's Vireo	3
Vireonidae	Warbling Vireo	3
Parulinae	Tennessee Warbler	3
Parulinae	American Redstart	3
Parulinae	MacGillivray's Warbler	3
Thraupinae	Western Tanager	3
Cardinalinae	Indigo Bunting	3
Cardinalinae	Dickcissel	3
Cardinalinae	Rose-breasted Grosbeak	3
Icterinae	Orchard Oriole	3
Icterinae	Northern Oriole	3
WINTERS FROM U.S. SOUTHWARD		
Cathartidae	Turkey Vulture	4
Accipitridae	Osprey	4
Accipitridae	Northern Harrier	4
Accipitridae	Sharp-shinned Hawk	4
Accipitridae	Cooper's Hawk	4
Accipitridae	Red-tailed Hawk	4
Falconidae	American Kestrel	4
Falconidae	Merlin	4
Falconidae	Peregrine Falcon	4
Charadriidae	Killdeer	4
Scolopacidae	Willet	4
Scolopacidae	Spotted Sandpiper	4
Scolopacidae	Marbled Godwit	4
Columbidae	Mourning Dove	4
Strigidae	Burrowing Owl	4
Apodidae	White-throated Swift	4
Trochilidae	Ruby-throated Hummingbird	4
Alcedinidae	Belted Kingfisher	4
Picidae	Yellow-bellied Sapsucker	4
Picidae	Red-naped Sapsucker	4
Tyrannidae	Hammond's Flycatcher	4
Hirundinidae	Tree Swallow	4
Hirundinidae	Violet-green Swallow	4
Hirundinidae	Northern Rough-winged Swallow	4
Sylviinae	Ruby-crowned Kinglet	4
Muscicapinae	Hermit Thrush	4

Table 2. (continued)

Family/Subfamily	Species	Wintering Area
Mimidae	Gray Catbird	4
Motacillidae	American Pipit	4
Bombycillidae	Cedar Waxwing	4
Vireonidae	Solitary Vireo	4
Vireonidae	Yellow-throated Vireo	4
Parulinae	Orange-crowned Warbler	4
Parulinae	Nashville Warbler	4
Parulinae	Yellow Warbler	4
Parulinae	Yellow-rumped Warbler	4
Parulinae	Townsend's Warbler	4
Parulinae	Black-and-white Warbler	4
Parulinae	Ovenbird	4
Parulinae	Northern Waterthrush	4
Parulinae	Common Yellowthroat	4
Parulinae	Wilson's Warbler	4
Parulinae	Yellow-breasted Chat	4
Emberizinae	Savannah Sparrow	4
Emberizinae	Grasshopper Sparrow	4
Emberizinae	Lincoln's Sparrow	4

WINTERS IN U.S. AND MEXICO

Family/Subfamily	Species	Wintering Area
Accipitridae	Northern Goshawk	5
Accipitridae	Ferruginous Hawk	5
Falconidae	Prairie Falcon	5
Charadriidae	Mountain Plover	5
Scolopacidae	Long-billed Curlew	5
Strigidae	Long-eared Owl	5
Strigidae	Short-eared Owl	5
Caprimulgidae	Common Poorwill	5
Picidae	Lewis' Woodpecker	5
Picidae	Williamson's Sapsucker	5
Tyrannidae	Dusky Flycatcher	5
Tyrannidae	Eastern Phoebe	5
Tyrannidae	Say's Phoebe	5
Troglodytidae	Rock Wren	5
Troglodytidae	House Wren	5
Troglodytidae	Sedge Wren	5
Troglodytidae	Marsh Wren	5
Muscicapinae	Mountain Bluebird	5
Muscicapinae	Townsend's Solitaire	5
Mimidae	Sage Thrasher	5
Motacillidae	Sprague's Pipit	5
Laniidae	Loggerhead Shrike	5
Cardinalinae	Lazuli Bunting	5
Cardinalinae	Black-headed Grosbeak	5
Emberizinae	Green-tailed Towhee	5
Emberizinae	Chipping Sparrow	5
Emberizinae	Clay-colored Sparrow	5
Emberizinae	Brewer's Sparrow	5
Emberizinae	Vesper Sparrow	5
Emberizinae	Lark Sparrow	5
Emberizinae	Sage Sparrow	5
Emberizinae	Lark Bunting	5
Emberizinae	Baird's Sparrow	5
Emberizinae	Fox Sparrow	5
Emberizinae	Swamp Sparrow	5
Emberizinae	White-crowned Sparrow	5
Emberizinae	McCown's Longspur	5
Emberizinae	Chestnut-collared Longspur	5
Icterinae	Yellow-headed Blackbird	5
Fringillidae	Cassin's Finch	5
Fringillidae	American Goldfinch	5

RESIDENT IN U.S. & WINTERS SOUTHWARD

Family/Subfamily	Species	Wintering Area
Accipitridae	Golden Eagle	6
Alaudidae	Horned Lark	6
Certhiidae	Brown Creeper	6
Muscicapinae	Eastern Bluebird	6
Muscicapinae	Western Bluebird	6
Muscicapinae	American Robin	6
Mimidae	Northern Mockingbird	6
Emberizinae	Rufous-sided Towhee	6
Emberizinae	Song Sparrow	6
Emberizinae	Dark-eyed Junco	6
Icterinae	Red-winged Blackbird	6
Icterinae	Western Meadowlark	6
Icterinae	Brewer's Blackbird	6
Fringillidae	Pine Siskin	6

SPECIES ACCOUNTS

TURKEY VULTURE

Cathartes aura
Cathartidae

Summer Resident (may occasionally winter in westernmost portion of region)

WINTERING AREA: 4

HABITAT REQUIREMENTS: Open areas from plains into the mountains, generally associated with drier brushy woodlands and grasslands. Usually nests on a cliff ledge or in a cliff niche, occasionally sheltered in a cave or hollow stump, only rarely on a dead snag.

FEEDING: A scavenger par excellence that consumes a tremendous variety of carrion; if it's animal matter and dead, it's food. Prey is located by sight and by smell while soaring at varied heights.

STATUS AND MANAGEMENT: Numbers have decreased significantly in Montana and declined in Idaho, as well. Data are insufficient from the remainder of the region. For the West as a whole, there have been slight declines but considering the entire North American range, the picture is mixed with an apparent increase following declines due in part to widespread eggshell thinning attributable to pesticides ingested in contaminated prey. Adequacy of roost sites free from disturbance for this communally roosting species may be an important component of maintaining local populations. Favored communal roost sites in winter in eastern U.S. are composed of stands of large conifers (Thompson et al., 1990).

FURTHER READING: Arad et al., 1989; Clark and Ohmart, 1985; Coleman and Fraser, 1989; Prior, 1990; Wilbur and Jackson, 1983.

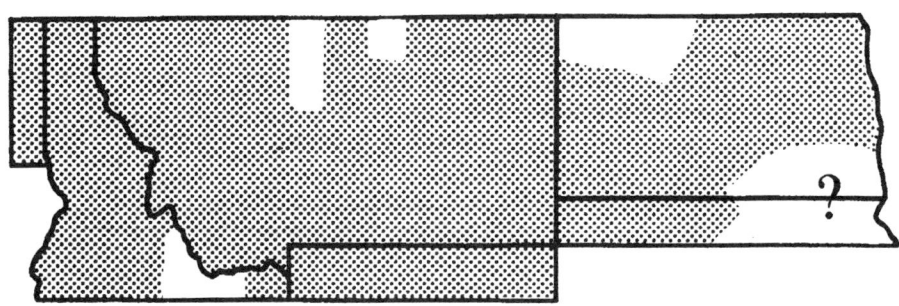

OSPREY

Pandion haliaetus
Accipitridae

Summer Resident

WINTERING AREA: 4

HABITAT REQUIREMENTS: Closely associated with rivers and lakes over a wide altitudinal range. Nests primarily in large trees, either live or dead but usually with broken tops; also nests on human-built structures, and occasionally on cliff ledges. Nests are often reused in subsequent years.

FEEDING: Nearly exclusively feeds on fish, captured by hovering over water and then diving (generally from 20–30 m) onto prey, plunging feet-first to grasp fish in their talons. Also captures fish Bald Eagle-style by snatching prey directly from water following low flight. Only rarely takes rodents, birds, or other small vertebrates, and crustaceans.

STATUS AND MANAGEMENT: Numbers are small but increasing significantly in Montana and increasing slightly in Idaho. Declined precipitously throughout North America as a result of pesticide-induced eggshell thinning, but has responded well to conservation efforts in many areas. Average dispersal distances of birds reared in the Northern Region are much greater than distances moved by eastern birds (Johnson and Melquist, 1991). Eggshell thinning and reduced hatching success resulting from pesticide contamination continue due to exposure in wintering areas outside of the U.S. and from contaminated prey consumed on the North American breeding grounds (Steidl et al., 1991). Readily nests on human-built nesting platform atop pole. Ospreys do not maintain large exclusive breeding territories but instead defend only the immediate area around the nest site, thus enabling unusually high nest densities if adequate supply of fish is present.

FURTHER READING: Bird, 1983; Edwards, 1989; Hagan and Walters, 1990; Johnsgard, 1990; Poole, 1989; Steidl and Griffin, 1991; Van Daele and Van Daele, 1982.

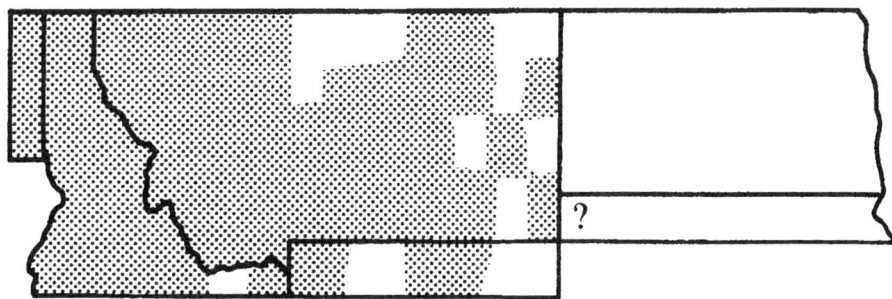

NORTHERN HARRIER

Circus cyaneus
Accipitridae

Accipitridae

Permanent Resident in western portion of region and Summer Resident from eastern Montana through North Dakota

WINTERING AREA: 4

HABITAT REQUIREMENTS: Prairies, open shrublands, wet meadows, marshes, croplands. Nests generally are on the ground but occasionally elevated in low shrubs, emergent vegetation, or low, dense riparian thickets. Avoids forested areas and most often found in the vicinity of water.

FEEDING: Characteristically forages by coursing low over the ground systematically searching for prey by sound as well as by sight. Feeds primarily on small mammals, especially voles, but will shift to fledgling passerines when available seasonally; also takes snakes, frogs, large insects, and even carrion.

STATUS AND MANAGEMENT: Montana populations have declined significantly, with Idaho showing a steady decline, as well. Until recently, North Dakota populations have been stable. Northern Harriers have declined steadily throughout their range until recent years when the magnitude of the decline increased significantly, with dramatic reductions in Wyoming and North Dakota in the most recent surveys. Nesting areas subject to heavy livestock grazing are particularly vulnerable, especially during dry years. Destruction of favored wetland habitats and eggshell thinning due to pesticide ingestion have been implicated in the long-term decline of populations.

FURTHER READING: Barnard et al., 1987; Hamerstrom, 1986; Johnsgard, 1990; Martin, 1987; Rice, 1982; Simmons, 1988; Temeles, 1989.

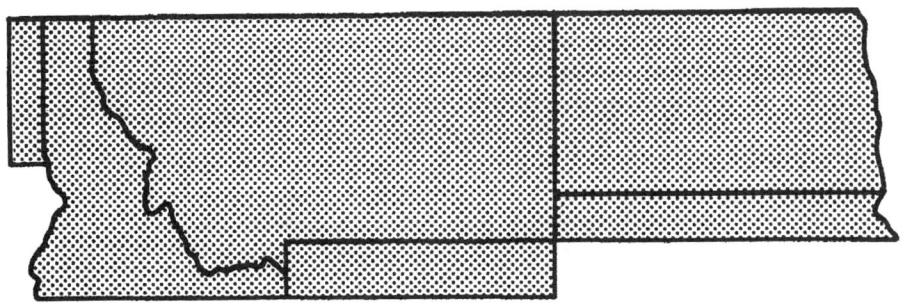

SHARP-SHINNED HAWK

Accipiter striatus
Accipitridae

Permanent Resident throughout region but more commonly a Summer Resident from central Montana through North Dakota

WINTERING AREA: 4

HABITAT REQUIREMENTS: Favors coniferous or mixed forests and open woodlands. Nests in trees, usually conifers, although aspens and riparian deciduous trees are not uncommonly used in the region. Rarely found outside of woodland or forested areas.

FEEDING: A specialist on small birds, only rarely taking small mammals, frogs, lizards, and large insects. Pursues prey with a quick, often erratic dash among the trees and seizes birds in flight or from their perches.

STATUS AND MANAGEMENT: Sample sizes for BBS data are rather limited but nevertheless indicate highly significant declines in Montana and North Dakota breeding populations. Nationally, significant declines in breeding populations occurred during the 1970s (especially in the East), attributable in part to nesting failure due to eggshell thinning. More recently, numbers appear to have stabilized and even increased in some areas. Tends to nest in dense, even-aged coniferous forests 25 to 60 years old (Reynolds et al., 1982), especially with large crown canopy volumes (Moore and Henny, 1983), usually near water (Reynolds et al., 1982). Sensitive to forest fragmentation in old-growth Douglas fir/ponderosa pine (Aney, 1984; Rosenberg and Raphael, 1986). Sharp-shinned Hawks are highly vulnerable to timber harvest because of their specific nesting habitat requirements; active and prospective nest sites should not be precommercially or commercially thinned (Reynolds, 1983).

FURTHER READING: Henny et al., 1985; Johnsgard, 1990; Kerlinger and Lehrer, 1982; Platt, 1976; Reynolds and Meslow, 1984.

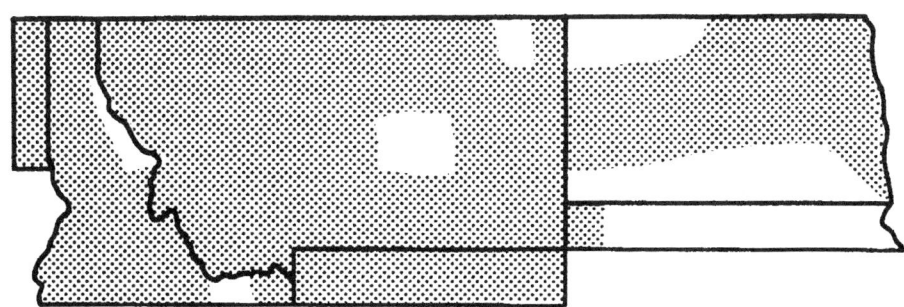

COOPER'S HAWK

Accipiter cooperii
Accipitridae

Accipitridae

Permanent Resident throughout region but more commonly a Summer Resident from eastern Montana through North Dakota

WINTERING AREA: 4

HABITAT REQUIREMENTS: Favors deciduous or mixed mature forest or woodland, as well as open and riparian woodlands. Nests in deciduous trees, especially large aspens, less commonly in conifers, often close to water.

FEEDING: Feeds extensively on young songbirds but takes proportionately more mammalian prey than the Sharp-shinned Hawk; only rarely includes amphibians in diet but reptiles may be important component in more arid areas. Hunts with rapid, relatively low flight among trees to flush avian prey and capture them in air.

STATUS AND MANAGEMENT: BBS data are scant but numbers appear to be declining throughout the region, especially in North Dakota. Nationally, Cooper's Hawks have declined significantly in association with pesticide-induced eggshell thinning. Populations largely stabilized since the early 1970s, although numbers appear to be declining in most recent survey years. Officially listed as Vulnerable in Canada. When nesting in coniferous forest or mixed forest, tends to nest in older even-aged stands; favors north-facing slopes for nest placement (Reynolds et al., 1982). Sensitive to forest fragmentation in old-growth ponderosa pine/Douglas fir (Aney, 1984). Cooper's Hawks are highly vulnerable to timber harvest because of their specific nesting habitat requirements; active and prospective nest sites should not be precommercially or commercially thinned (Reynolds, 1983).

FURTHER READING: Asay, 1987; Bielefeldt et al., 1992; Henny et al., 1985; Johnsgard, 1990; Kennedy and Johnson, 1986; Reynolds and Meslow, 1984; Rosenfield et al., 1991.

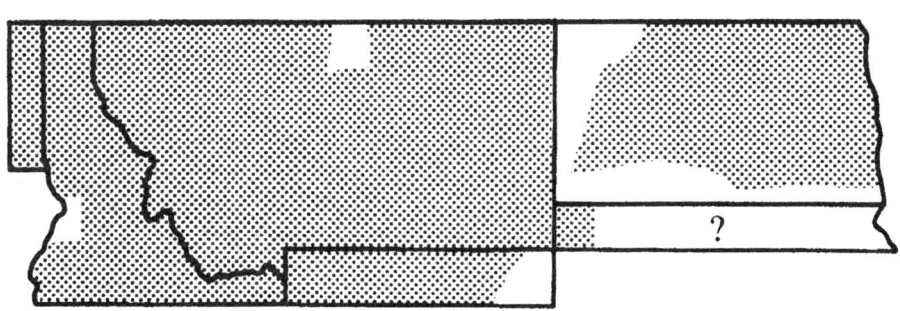

NORTHERN GOSHAWK

Accipiter gentilis
Accipitridae

Permanent Resident except in North Dakota and easternmost Montana where it is a Winter Resident only.

WINTERING AREA: 5

HABITAT REQUIREMENTS: Coniferous or mostly coniferous forest and open woodland with significant old-growth component. Nests in large, old-growth conifers or sometimes aspens, especially near water. Nests are often repaired and reused.

FEEDING: Diet composed of only slightly more birds than mammals. Avian prey are primarily gallinaceous birds and larger songbirds; mammalian prey are mostly ground squirrels and lagomorphs; only rarely capture snakes and large insects. Females take considerably larger prey than males, with little overlap between the two. Attacks prey from perch to a greater extent than the other accipiters, but often forages in a low patrolling flight to flush avian prey.

STATUS AND MANAGEMENT: Virtually absent from the BBS database. Numbers apparently have declined precipitously in recent years in association with the disappearance and fragmentation of old-growth coniferous forests. Recently listed as a candidate species for Threatened and Endangered status throughout its range in the U.S. Closely associated during breeding season with dense old-growth or mature coniferous forests having multilayered canopies with average canopy closures of 80 to 88 percent; favors nesting sites near water on moderate north-facing slopes (Hayward and Escano, 1989; Hejl and Woods, 1991; Mannan and Meslow, 1984; Reynolds et al., 1982). Home ranges are quite large, thus requiring extensive forest stands. Northern Goshawks are highly vulnerable to most timber management activities because of their specific nesting habitat requirements; active and prospective nest sites should not be precommercially or commercially thinned (Reynolds, 1983). Populations irrupt southward on a ten-year cycle in association with crash of prey populations in northern areas, although this pattern is most pronounced in eastern portion of range.

FURTHER READING: Henny et al., 1985; Johnsgard, 1990; Mueller et al., 1977; Reynolds and Meslow, 1984; Sherrod, 1978.

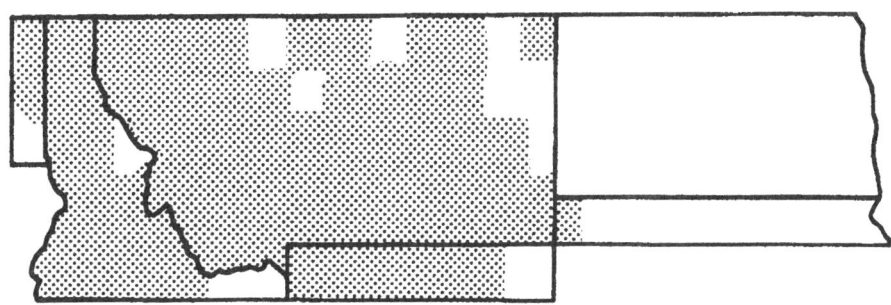

BROAD-WINGED HAWK

Buteo platypterus
Accipitridae

Accipitridae

Summer Resident

WINTERING AREA: 3

HABITAT REQUIREMENTS: Mature deciduous or mixed forest, especially associated with flood plains of major watercourses and lakes. Usually nests in deciduous trees.

FEEDING: Primary prey are small mammals (mice, voles, lagomorphs, squirrels, shrews) but fledglings and nestlings of smaller birds are often important components of diet; also takes amphibians (especially frogs), reptiles, and insects. Forages by dropping onto prey from an elevated perch.

STATUS AND MANAGEMENT: Primarily an eastern species, the Broad-winged Hawk does not appear in BBS data for the region due to its rarity here. Nationally, populations appear relatively stable with modest increases in some regions and slight decreases in others. Suitable nesting habitat is limited primarily to eastern North Dakota. Extremely vulnerable to tropical deforestation (Morton, 1992).

FURTHER READING: Crocoll and Parker, 1989; Fitch, 1974; Johnsgard, 1990; Matray, 1974; Rosenfield, 1984.

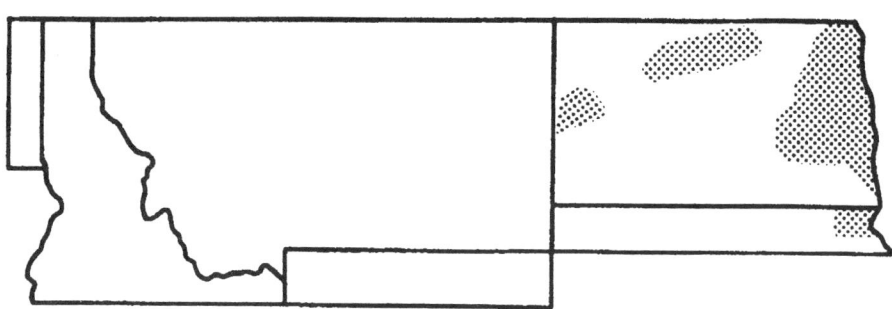

SWAINSON'S HAWK

Buteo swainsoni
Accipitridae

Summer Resident

WINTERING AREA: 1

HABITAT REQUIREMENTS: Shrubsteppe, prairies, open woodland, shelterbelts, cultivated land with scattered trees. Nests typically in isolated deciduous trees, much less commonly in shrub or conifer, on cliff ledge or ground in treeless areas. Often refurbish and use abandoned nests of corvids and other *Buteo* species.

FEEDING: Rodents (especially ground squirrels and pocket gophers) comprise most of diet but also consumes lagomorphs and a wide variety of small vertebrates (including birds), as well as beetles and grasshoppers, which are sometimes taken in large numbers. Forages primarily by soaring at moderate heights and stooping onto terrestrial prey but also frequently swoops down onto prey from elevated perches.

STATUS AND MANAGEMENT: Numbers have increased significantly in North Dakota and Montana, and remain fairly steady in Idaho. In the West as a whole, numbers appear to have increased slightly, with most of the increase occurring in the northern portion of the range. Egg contamination with low to moderate levels of organochlorine pesticides, PCBs, and mercury has been found within the region (Stendell et al., 1988). Often interspecifically territorial with Red-tailed Hawks when nesting, Swainson's Hawks are more tolerant of landscapes having agricultural habitats interspersed with grasslands than are Ferruginous Hawks (Gilmer and Stewart, 1984; Schmutz, 1989).

FURTHER READING: Bechard et al., 1990; Dunkle, 1977; Fitzner, 1978; Houston et al., 1991; Johnsgard, 1990; Kirkley and Gessaman, 1990a, b; Restani, 1991.

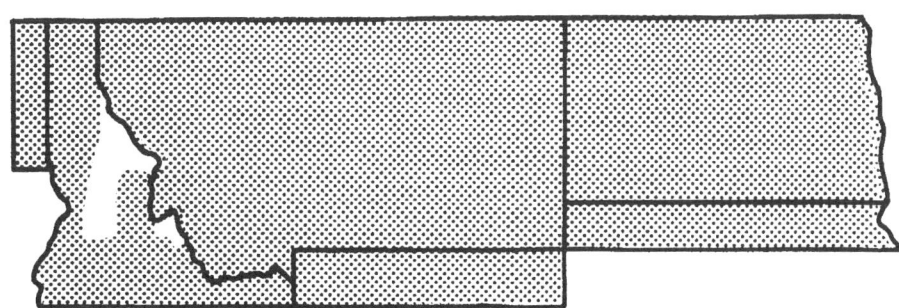

RED-TAILED HAWK

Buteo jamaicensis
Accipitridae

Permanent Resident in western portion of region, mostly Summer Resident from north-central Montana through North Dakota

WINTERING AREA: 4

HABITAT REQUIREMENTS: Wide variety of open woodland and other open country with scattered trees or isolated woodlots. Prefers to nest in large deciduous trees (and ponderosa pines) but will use cliff ledge or powerline tower where suitable trees are unavailable. May alternate among several nests over time.

FEEDING: Rodents and lagomorphs predominate but also take birds, reptiles (especially snakes), large insects, amphibians, crayfish, fish, and even carrion. Forages primarily by soaring at moderate heights and stooping onto terrestrial prey but also frequently swoops down onto prey from elevated perches.

STATUS AND MANAGEMENT: Populations appear to be increasing significantly in Montana, and relatively stable elsewhere in the region. Overall, populations have posted a small but significant gain, especially in the West, but numbers are down significantly in some areas (including the Canadian prairie provinces) in the most recent survey years. Often interspecifically territorial with Swainson's Hawks when nesting, especially when favored prey densities are sparse. This is the most common and widespread species of *Buteo* in North America, and the most tolerant of human disturbance. Increases in recent years may have been at the expense of Swainson's and Ferruginous Hawks, which are less tolerant of habitat alteration and human disturbance.

FURTHER READING: Andersen, 1990; Gilmer et al., 1983; Janes, 1984a, b; Johnsgard, 1990; Kirkley and Gessaman, 1990b; Restani, 1991; Rothfels and Lein, 1983.

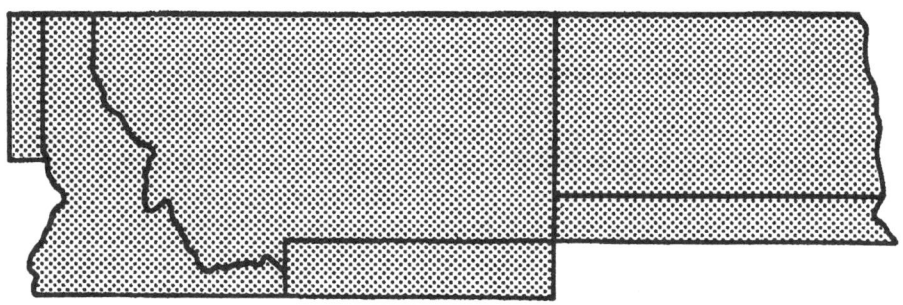

FERRUGINOUS HAWK

Buteo regalis
Accipitridae

Summer Resident

WINTERING AREA: 5

HABITAT REQUIREMENTS: Dry open country, especially native prairies, but also shrubsteppe, plains, and badlands. Preferred nest site is in a tree (deciduous in eastern portion of region, coniferous in western portion), but also nests on cliff ledge, atop rock outcrop, in deciduous shrub, or on elevated ground, and on human-built structures, including haystacks. Elevated nests are often used perennially.

FEEDING: Primarily small mammals, especially ground squirrels, jackrabbits, and pocket gophers, but also takes birds, reptiles, and large insects. Foraging behaviors are varied: often attacks prey by a low, rapid flight and pursuit; also swoops onto prey from elevated perches or soars at varied heights in search of prey, on which it often drops after pulling up and hovering in flight.

STATUS AND MANAGEMENT: Numbers appear to have increased significantly in Montana, slightly in North Dakota, and decreased somewhat in Idaho. Populations appear to be stable overall following marked declines in many areas beginning in the late 1940s, although the Ferruginous Hawk is still uncommon in many parts of its breeding range. Listed as Threatened in Canada and as a candidate for Threatened status in the U.S. Loss of suitable habitat to agricultural conversion and overgrazing by livestock is the primary problem faced by the species. Egg contamination with low to moderate levels of organochlorine pesticides, PCBs, and mercury has been found within the region (Stendell et al., 1988). Optimal breeding habitat consists of extensive ungrazed or lightly grazed prairie grassland with scattered elevated nesting sites (Ensign, 1983); cultivated croplands cannot sustain populations (Gilmer and Stewart, 1983). Powerline towers providing secure platforms are used extensively as nest sites in North Dakota (Gilmer and Stewart, 1983). Nesting productivity often varies in synchrony with jackrabbit or ground squirrel abundance (Lokemoen and Duebbert, 1976).

FURTHER READING: Blair, 1978; Fitzner et al., 1977; Gilmer et al., 1985; Johnsgard, 1990; Restani, 1991; Smith et al., 1981; Woffinden and Murphy, 1989.

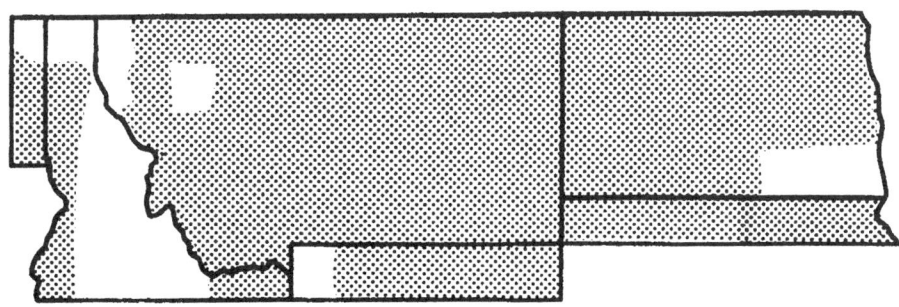

GOLDEN EAGLE

Aquila chrysaetos
Accipitridae

Permanent Resident except in eastern portion of region (central and eastern North Dakota) where it is a Winter Resident

WINTERING AREA: 6

HABITAT REQUIREMENTS: Open habitats in mountains and hill country most commonly but also prairies and other grasslands. Nests on cliff ledge or in cliff niche, less commonly in large tree. Pairs often alternate among favored nest sites, which may be used over many years.

FEEDING: Predominant prey are jackrabbits, but other mammals (especially ground squirrels), birds, reptiles, and even carrion are consumed. Forages by soaring widely and by swooping down onto prey from the air or from an elevated perch. Pair occasionally hunts in tandem. Requires thermals or topographically-induced updrafts to gain sufficient altitude for energetically efficient hunting.

STATUS AND MANAGEMENT: Slight declines are apparent in Idaho and Montana, with relatively steady numbers in North Dakota. Fewer than 100 nesting pairs were estimated for all of western North Dakota in the mid-1980s, with the recommendation that maintenance of suitable habitat should be a management priority (Allen 1986, 1987). Populations in the West generally have stabilized in most areas following slight but sustained local declines, especially where encroachment of suburban sprawl and agricultural conversion of prairie habitat have occurred. Widely persecuted in the past for depredations on livestock (primarily lambs)—which often occur as a result of poor livestock management practices (Brown and Watson, 1964) in association with population crashes of primary prey species (especially jackrabbits) and limited availability of alternate prey (Matchett and O'Gara, 1987). Numbers continue to be impacted adversely by inadvertent powerline electrocutions, illegal shooting, and death from ingestion of poisoned carcasses intended for coyotes. Lead poisoning also may be an important and underestimated problem in Golden Eagles (Craig et al., 1990).

FURTHER READING: Beecham and Kochert, 1975; Collopy, 1984; Johnsgard, 1990; Steenhof et al., 1983.

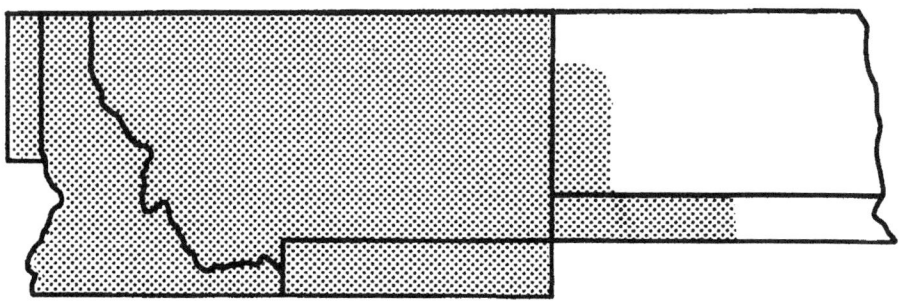

AMERICAN KESTREL

Falco sparverius
Falconidae

Summer Resident from central Montana eastward; Permanent Resident elsewhere in region

WINTERING AREA: 4

HABITAT REQUIREMENTS: Open country with or without scattered trees, including prairie, grasslands, mountain meadows, shrubsteppe, riparian woodland, and agricultural areas. Nests in cavities in trees, buildings, or niches in cliffs or rocks; does not excavate its own cavity.

FEEDING: Highly insectivorous, feeding heavily on grasshoppers, but also takes a wide variety of small vertebrates, especially rodents and birds, which together comprise a greater proportion of diet biomass than do insects. Hovers and pounces on ter-restrial prey or swoops down from elevated perch.

STATUS AND MANAGEMENT: Numbers have increased throughout the region, significantly so in Montana. Populations appear stable in North America overall, but recent significant declines have occurred in scattered areas of the Midwest and Southeast, and numbers are down slightly in the West when viewed as a whole. Breeding territory size has been shown to vary inversely with small rodent densities (Gard and Bird, 1990). Availability of suitable cavities for nest sites may depend heavily on Northern Flickers and other large woodpeckers in some areas. American Kestrels readily use nest boxes in the region (Wheeler, 1992).

FURTHER READING: Cade, 1982; Johnsgard, 1990; Sedgwick and Knopf, 1990; Toland, 1987; Varland et al., 1991.

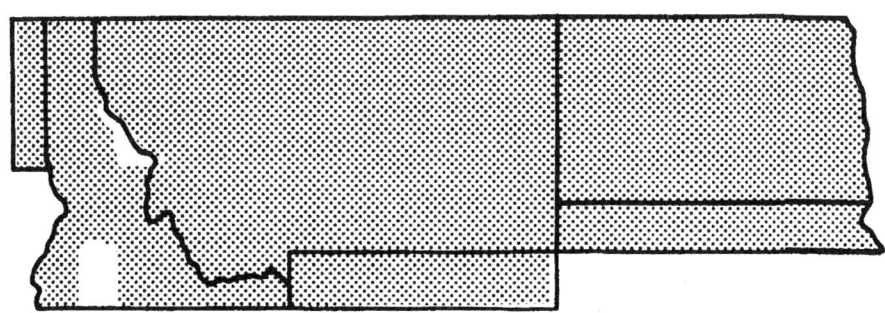

MERLIN

Falco columbarius
Falconidae

Summer Resident but occurs sporadically as a Winter Resident throughout region

WINTERING AREA: 4

HABITAT REQUIREMENTS: Open habitats at all elevations, often associated with grasslands and shrubsteppe. For nesting, relies heavily on abandoned magpie nests in large coniferous trees (most often ponderosa pines) in open woodland close to prairie or other grassland habitat. In other areas, Merlins are known to use nests of corvids (sometimes hawks) in deciduous trees or shrubs in open woodlands; where tree sites are unavailable, nesting may occur on the ground beneath shrubs or on cliff ledges, or rarely in cavities of standing snags.

FEEDING: Diet is composed almost wholly of songbirds found in open habitats, which are captured most often by direct, very fast flight rather than by stooping on flying prey from above as do most larger falcons. Merlins also will hunt from elevated perches and by typical falcon-stoop from above. Very small portion of diet (<10%) consists of smaller rodents and large insects (especially dragonflies) but only rarely includes other small vertebrates.

STATUS AND MANAGEMENT: Although sample sizes are small, BBS data indicate a highly significant decline in numbers in Idaho and Montana, with no data from North Dakota. Widespread eggshell thinning occurred in North American populations through the 1970s concurrent with widespread decline in numbers. Breeding range contraction and population decline have been attributed to pesticide contamination and destruction of native prairie habitats; pesticide contamination associated with significantly reduced eggshell thicknesses continues to be of concern in the region (Becker and Sieg, 1987a). A rather uncommon breeding bird in the region with most breeding and wintering records from Montana, in contrast to only a handful of nesting records known from widely scattered locations in North Dakota. Home ranges are very large, and preferred hunting habitats consist of interspersed areas of grassland and shrubsteppe (Becker and Sieg, 1987b); removal of shrubs or conversion of sagebrush/grassland to agricultural cropland markedly reduces quality of foraging habitat (Becker and Sieg, 1987b).

FURTHER READING: Becker and Sieg, 1985; Buchanan et al., 1988; Cade, 1982; Johnsgard, 1990; Schmutz et al., 1991; Sieg and Becker, 1990; Sodhi et al., 1991; Warkentin et al., 1990.

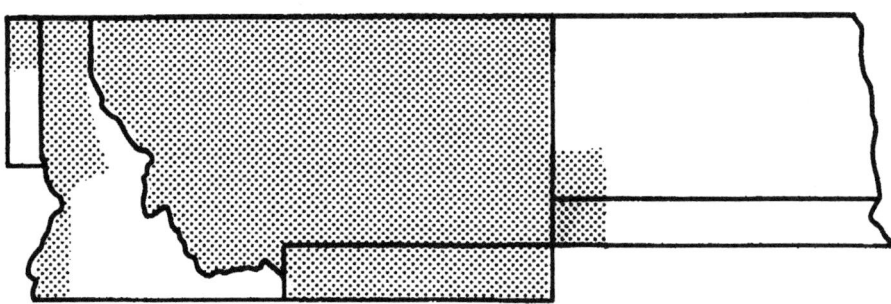

PEREGRINE FALCON

Falco peregrinus
Falconidae

Summer Resident

WINTERING AREA: 4

HABITAT REQUIREMENTS: A cliff-nesting falcon of open habitats, most commonly associated with open woodlands, but found from prairie to alpine tundra. Cliff nests are used perennially. Rarely nests in trees using abandoned corvid or hawk nest or cavity.

FEEDING: A specialist on avian prey, especially doves, waterfowl, shorebirds, and a wide variety of passerines. Strikes prey in air following extremely fast, low flight to flush birds from ground, or stoops on prey from above, but also attacks passing birds from stationary, elevated perch. Opportunistically includes small mammals in diet.

STATUS AND MANAGEMENT: No BBS data exist for Peregrines in the region due to the species' rarity; what few data there are, indicate a slight declining trend in the West as a whole. The subspecies that breeds in our region is federally listed as Endangered, but status is now under review as a result of apparent improvement in overall population numbers. The arctic-breeding subspecies is federally listed as Threatened and a third subspecies is listed as Vulnerable in Canada. Extirpated from much of its former range as a result of habitat destruction and pesticide contamination that produced extensive eggshell thinning (which continues to be a problem in some parts of the U.S. [e.g., see Steidl et al., 1991]); now being reintroduced in portions of its previous range, with more than 400 captive-reared birds released in the Rockies since 1976.

FURTHER READING: Cade, 1982; Cade et al., 1988; Court et al., 1988; Craig, 1986; Hunter et al., 1988; Johnsgard, 1990; Ratcliffe, 1980; Schmutz et al., 1991; Temple, 1978.

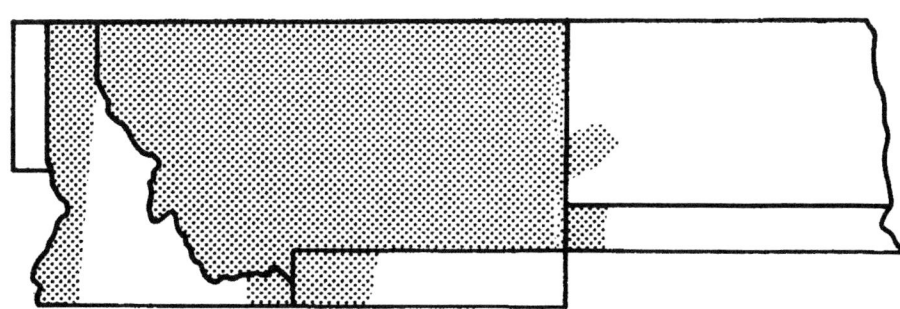

PRAIRIE FALCON

Falco mexicanus
Falconidae

Summer Resident or Permanent Resident

WINTERING AREA: 5

HABITAT REQUIREMENTS: A cliff-nesting species of open habitats from prairie to alpine tundra; most commonly associated with arid habitats such as shortgrass prairie and shrubsteppe. Cliff nests are regularly reused. Rarely nests in abandoned raven nest in tree or on power-transmission tower.

FEEDING: Mostly birds but also preys substantially on rodents (which often comprise most of diet in some areas), lagomorphs, and other small mammals, with reptiles and large insects accounting for only 5 to 10 percent of diet. Flushes prey with low, searching flight, stoops on avian prey from high soaring position, and searches for prey from elevated perch.

STATUS AND MANAGEMENT: Numbers appear to be relatively stable throughout the region, as also appears to be the case elsewhere, although there is a suggestion of slight but sustained decline in numbers overall; BBS data are rather scant. The breeding population in North Dakota has been estimated at only 125 pairs (Allen, 1987). Pesticide contamination has led to eggshell thinning and mercury poisoning, but the species has not experienced the extensive range reduction seen in Peregrine Falcons; in Montana, significant eggshell thinning correlated with elevated organochlorine residues has been associated with increased land areas devoted to alfalfa production (Leedy, 1972). Permanent residents from higher elevations often move to lower elevations during the winter where they hunt over grain fields and prey heavily on Horned Larks (Beauvais et al., 1992).

FURTHER READING: Allen et al., 1986; Cade, 1982; Holthuijzen, 1990; Johnsgard, 1990; Schmutz et al., 1991; Squires et al., 1989, 1993.

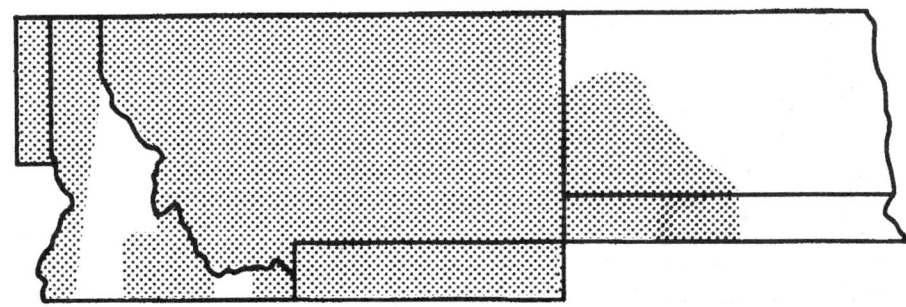

MOUNTAIN PLOVER

Charadrius montanus
Charadriidae

Summer Resident

WINTERING AREA: 5

HABITAT REQUIREMENTS: Shortgrass prairie, drier open shrubland. Nests on ground, often far from water in arid or semiarid conditions.

FEEDING: Forages on the ground, nearly exclusively on insects (especially beetles and grasshoppers), by picking items from the substrate.

STATUS AND MANAGEMENT: BBS data are inadequate to project population trends. North Dakota population has been extirpated. Listed as Endangered in Canada and as a candidate species for federal Threatened and Endangered status in the U.S. Range has contracted severely as prairie habitats have been converted to agriculture. Nesting Mountain Plovers prefer shortgrass prairie areas that are flat, moderately grazed, and provide short vegetation, hence most significant populations remaining in the northern region nest preferentially in prairie dog towns (Knowles et al., 1982; Olson, 1984); prairie dogs provide an optimum grazing pressure to maintain (but not eliminate) short-stature vegetation while selectively eliminating sagebrush (Olson, 1984). Control efforts aimed at prairie dogs will severely impact Mountain Plover populations in the region. Migrating birds often found in small flocks on plowed fields.

FURTHER READING: Graul, 1975; Graul and Webster, 1976; Knowles et al., 1982; Olson-Edge and Edge, 1987; Wershler, 1987.

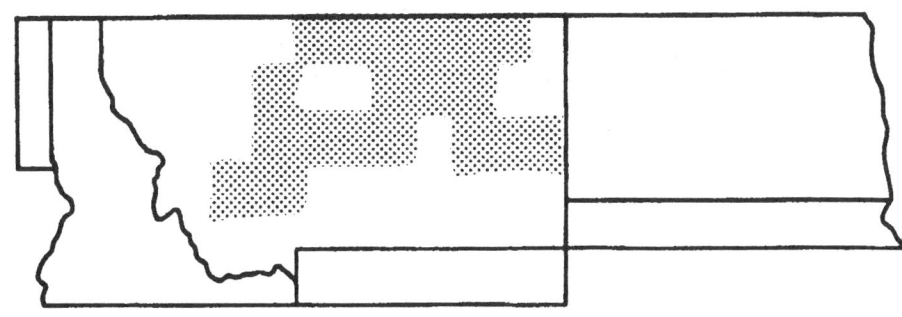

KILLDEER

Charadrius vociferus
Charadriidae

Summer Resident (Permanent Resident in some areas)

WINTERING AREA: 4

HABITAT REQUIREMENTS: Fields, meadows, pastures, freshwater margins; widely distributed in open habitats, most commonly at lower elevations. Nests on ground, usually with little cover, often in upland areas well away from water.

FEEDING: A ground forager that selectively picks food items from the substrate. Insects comprise about 75 percent of diet, with the remainder consisting of a wide variety of other invertebrates and occasional seeds of forbs and grasses.

STATUS AND MANAGEMENT: Regionally, numbers have declined slightly, although numbers appear steady over the entire North American range taken as a whole. Favors open, heavily grazed areas of prairie or pasture for nesting but may suffer decreased nesting success due to trampling of eggs and young by livestock. Migrating and wintering birds associate more closely with wetland habitats.

FURTHER READING: Brunton, 1988; Powell and Cuthbert, 1993.

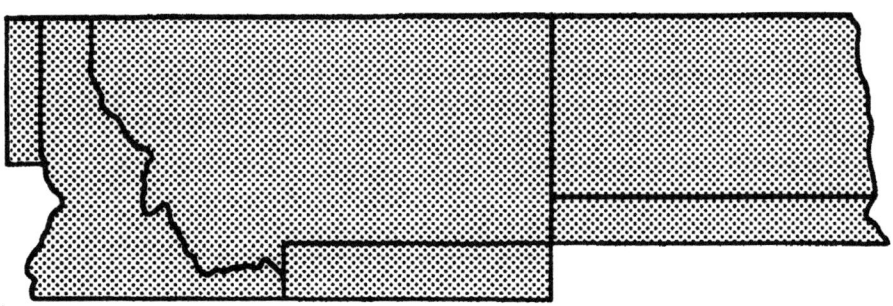

WILLET

Catoptrophorus semipalmatus
Scolopacidae

Summer Resident

WINTERING AREA: 4

HABITAT REQUIREMENTS: Marshy lake margins and adjacent prairie, meadow, or other grassy area. Nests on ground, generally within 200 m of water. Nesting areas similar to those favored by Marbled Godwits.

FEEDING: Feeds primarily by probing for aquatic invertebrates in soft, usually moist, substrates but also pick items from ground or water surface.

STATUS AND MANAGEMENT: Appears to be declining both regionally and nationally, although long-term data are scant. Nests preferentially in native prairie rather than in grassland seeded with exotic species (Kantrud and Higgins, 1992).

FURTHER READING: Howe, 1982; Ryan and Renken, 1987; Sordahl, 1979; Wilcox, 1980.

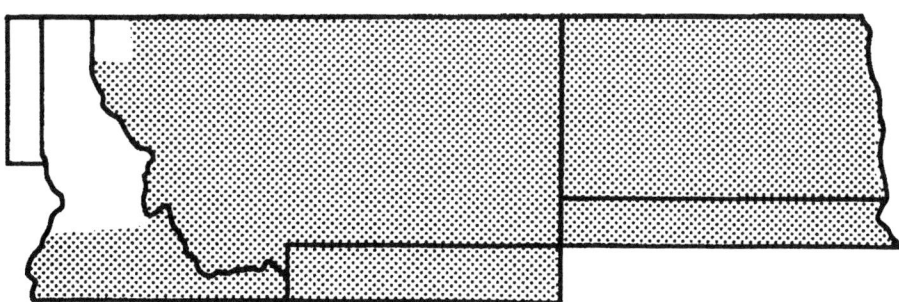

57

SPOTTED SANDPIPER

Actitis macularia
Scolopacidae

Summer Resident

Wintering Area: 4

HABITAT REQUIREMENTS: Always in the vicinity of water from ponds and lakes to (more typically) swiftly flowing rivers and streams with rocky bottoms. Occurs across the complete altitudinal spectrum in the region, especially favoring montane areas. Nests on the ground, usually near water but occasionally at some distance in vegetation ranging from grass to shrubs and trees.

FEEDING: Forages on the ground adeptly capturing flying insects, which comprise most of diet. Also takes other terrestrial and aquatic invertebrates from the ground or water's surface.

STATUS AND MANAGEMENT: BBS data are insufficient for the region but the species appears to be declining nationally, especially in the West. Continued degradation of riparian habitats may be an important contributor to western declines.

FURTHER READING: Alberico et al., 1991; Oring and Lank, 1986; Pickett et al., 1988.

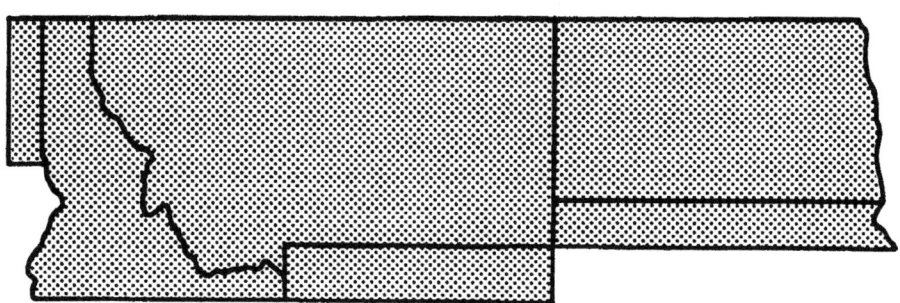

UPLAND SANDPIPER

Bartramia longicauda
Scolopacidae

Summer Resident

WINTERING AREA: 1

HABITAT REQUIREMENTS: Associated most closely with prairie grasslands, but also found in wet and dry meadows, hayfields, and pastures, seldom near water. Nest is well concealed on the ground.

FEEDING: Feeds primarily on insects and other terrestrial invertebrates, occasionally taking waste grain. Forages by picking items from the substrate, only rarely wading to feed.

STATUS AND MANAGEMENT: Loss of suitable breeding habitat has led to significant decline in eastern North America and contraction of range westward where intensive livestock grazing and conversion of prairie grassland to agriculture have impacted numbers. Loss of habitat to agriculture on the South American wintering grounds also has been implicated in population decline. Numbers appear to have stabilized in recent years throughout the breeding range, showing a small but significant improvement in some areas over the past 25 years. On mixed-grass native prairie in North Dakota, the presence of livestock (even at low stocking densities) during the nesting season has a detrimental effect on sandpiper reproduction, and results in significantly lower nest densities (Bowen and Kruse, 1992).

FURTHER READING: Dale, 1984; Higgins and Kirsch, 1975: Kaiser, 1979; Kirsch and Higgins, 1976.

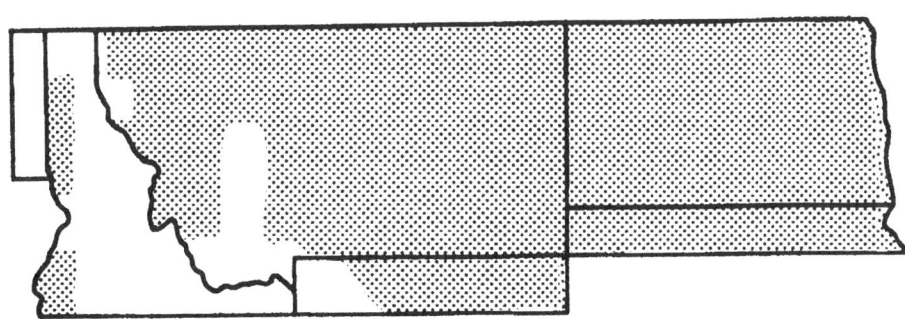

LONG-BILLED CURLEW

Numenius americanus
Scolopacidae

Summer Resident

WINTERING AREA: 5

HABITAT REQUIREMENTS: Shortgrass, grazed mixed-grass, and arid scrub prairies, meadows, not necessarily in association with water. Nests on ground, favoring well-drained, gravely soils; although nests are generally placed well away from water, curlews often forage in wetter areas.

FEEDING: Forages on the ground, picking items from the substrate or probing with its bill into soft substrates for concealed prey. Feeds primarily on insects but also takes a wide variety of other invertebrates, especially worms, crustaceans, and mollusks, as well as preying on amphibians and the eggs and nestlings of small birds. Berries also consumed, especially prior to fall migration.

STATUS AND MANAGEMENT: BBS data indicate a slight long-term decline but sample sizes are rather small. Suitable breeding habitat has been much reduced and breeding range continues to retrench westward. Populations appear also to have been impacted by organochlorine poisoning. Listed as a candidate species for federal Threatened and Endangered status.

FURTHER READING: Allen, 1980; Redmond and Jenni, 1982, 1986.

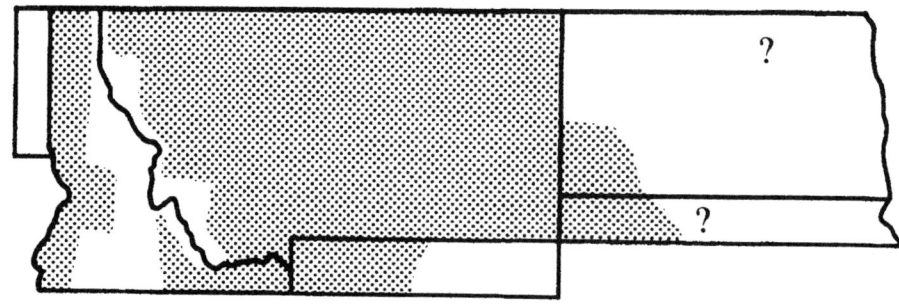

MARBLED GODWIT

Limosa fedoa
Scolopacidae

Summer Resident

WINTERING AREA: 4

HABITAT REQUIREMENTS: Shortgrass prairie, wet meadows and other low-lying wet areas. Nests on ground, often in grassy areas adjacent to wetlands, occasionally farther from water in prairie habitat but does not favor areas as dry as those used by nesting Long-billed Curlews.

FEEDING: Forages on the ground, generally probing into soft substrates in wet areas for concealed aquatic invertebrates; also forages by picking insects and other items from the ground surface.

STATUS AND MANAGEMENT: Numbers have declined regionally but data are largely lacking. Breeding range has contracted as favored habitats have been converted to agriculture or degraded by overgrazing. Preferentially nests in native prairie rather than in grassland seeded with exotic species (Kantrud and Higgins, 1992).

FURTHER READING: Gibson and Kessel, 1989; Ryan et al., 1984; Wishart and Sealy, 1980.

Scolopacidae

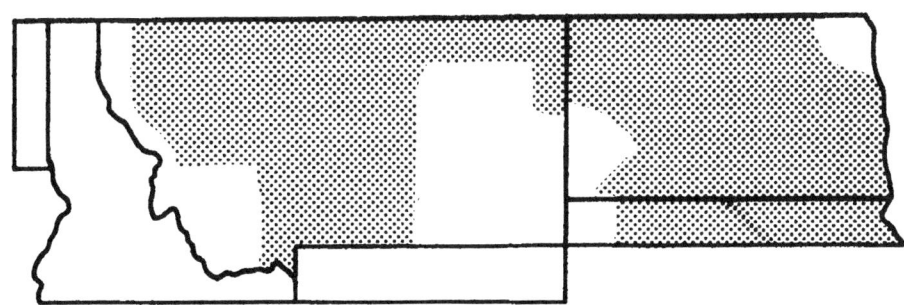

MOURNING DOVE

Zenaida macroura
Columbidae

Summer Resident but occasionally a Permanent Resident, especially in western portion of region

WINTERING AREA: 4

HABITAT REQUIREMENTS: Open woodlands (including riparian), montane meadows, cultivated lands with scattered trees and shrubs, arid grasslands near water. Nests in a wide variety of settings, most commonly in a deciduous tree but also in conifer, on abandoned tree nests of other birds, on the ground beneath shrubs in shrubsteppe and grassland habitats, and on human-built structures.

FEEDING: Forages on the ground taking seeds and waste grain.

STATUS AND MANAGEMENT: Populations appear stable in most of the region with significantly increasing numbers in North Dakota. On a continent-wide basis, numbers have been stable generally with marked recent increases in some regions, but there is also a small but significant declining trend for the West when viewed as a whole. A popular gamebird, also a regular visitor at backyard feeders and highly tolerant of human disturbance. Capable of producing up to six broods yearly.

FURTHER READING: Blockstein, 1989; Howe and Flake, 1989; Leopold et al., 1981; Losito et al., 1990; Walsberg and Schmidt, 1992.

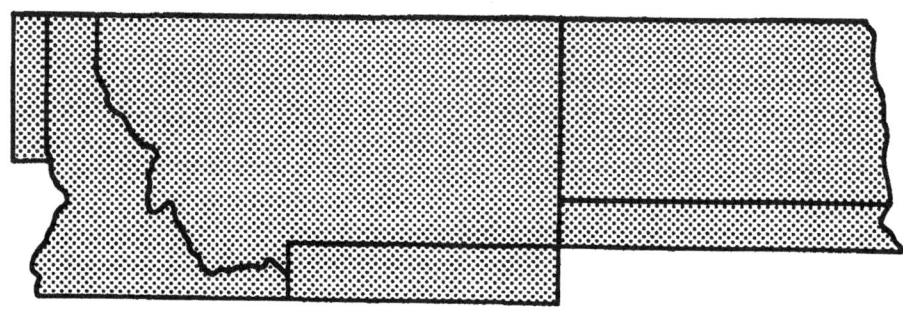

YELLOW-BILLED CUCKOO

Coccyzus americanus
Cuculidae

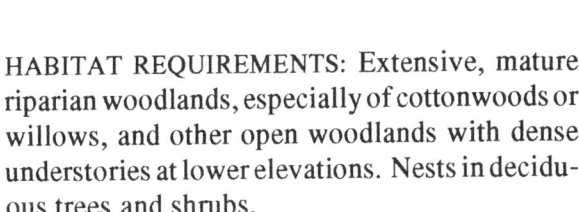

Summer Resident

WINTERING AREA: 1

HABITAT REQUIREMENTS: Extensive, mature riparian woodlands, especially of cottonwoods or willows, and other open woodlands with dense understories at lower elevations. Nests in deciduous trees and shrubs.

FEEDING: Primarily insects, especially caterpillars, but also takes bird eggs, small frogs and lizards, berries, and fruit. Mostly forages by gleaning from vegetation, but also hawks flying insects and hovers to glean from leaves.

STATUS AND MANAGEMENT: Extremely rare throughout the region, most commonly found in North Dakota; populations declining significantly throughout North American range. Numbers have declined precipitously in recent years to the point of extreme rarity throughout its previous range in the western U.S. Fragmentation and degradation of riparian woodlands by livestock, as well as drought, and prey scarcity caused by pesticides have all played roles in the species decline, although habitat loss and pesticides on the wintering grounds also are likely to have contributed to the problem. Yellow-billed Cuckoos in the West are extremely sensitive to fragmentation of mature cottonwood (and other riparian) habitat, requiring intact woodlands of at least 40 ha, preferably 80 ha or more (Laymon and Halterman, 1989). Listed as a candidate species for Threatened and Endangered status in the western U.S. Breeding productivity often varies in concert with outbreaks of caterpillars, cicadas, and other large insects. Highly vulnerable to tropical deforestation (Morton, 1992).

FURTHER READING: Anderson and Laymon, 1989; Laymon and Halterman, 1987.

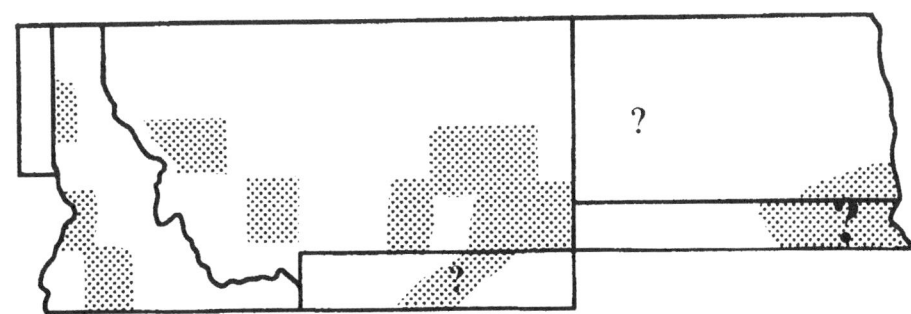

BLACK-BILLED CUCKOO

Coccyzus erythropthalmus
Cuculidae

Summer Resident

WINTERING AREA: 1

HABITAT REQUIREMENTS: Forest and open woodland of all types, generally at lower elevations, prairie thickets, and shelterbelts. Nests in trees or shrubs.

FEEDING: Primarily insects, especially caterpillars, but also takes bird eggs, small frogs and lizards, berries, and fruit. Forages in trees by gleaning arthropods from vegetation.

STATUS AND MANAGEMENT: Numbers appear to be low but stable in the region. Overall, populations appear to be stable throughout the species' range, although significant declines have occurred in the most recent survey years. Western populations taken as a whole display a slight declining trend. Nesting productivity often varies in concert with outbreaks of caterpillars, cicadas, and other large insects.

FURTHER READING: Sealy 1978, 1985.

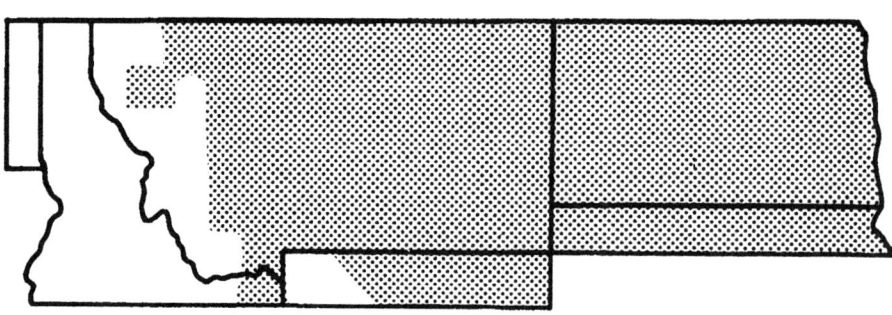

FLAMMULATED OWL

Otus flammeolus
Strigidae

Summer Resident

WINTERING AREA: 3

HABITAT REQUIREMENTS: Mature ponderosa pine forest, mixed ponderosa pine/Douglas fir forest, perhaps other montane coniferous forests to a lesser extent. Nests in woodpecker-excavated (sometimes natural) cavity in pine, larch, fir, and occasionally aspen.

FEEDING: Nearly exclusively invertebrates: insects, scorpions, spiders, and centipedes, only rarely taking small mammals, birds, or other small vertebrate prey. Captures flying insects in flight and gleans other prey from branches, trunks, and foliage of trees.

STATUS AND MANAGEMENT: Rather inconspicuous, Flammulated Owls do not appear in the regional BBS database, but may be more widespread in the western portion of the region than is currently known (Holt and Hillis, 1987). Flammulated Owls appear only rarely on BBS routes anywhere within their range, but the few data that exist indicate a significant decline. Closely associated with extensive stands of old-growth ponderosa pine and mixed old-growth ponderosa pine/Douglas fir, or grand fir (Mannan and Meslow, 1984; Bull et al., 1990), the species has disappeared from portions of its previous breeding range where such habitats have been highly fragmented and much diminished in extent. Flammulated Owls avoid cutover areas and forests younger than 100 years old (Reynolds and Linkhart, 1987b). Strictly nocturnal. Will use nest boxes.

FURTHER READING: Goggans, 1985; Howie and Ritcey, 1987; Johnsgard, 1988; Linkhart and Reynolds, 1987; McCallum and Gehlbach, 1988; Reynolds and Linkhart 1987a.

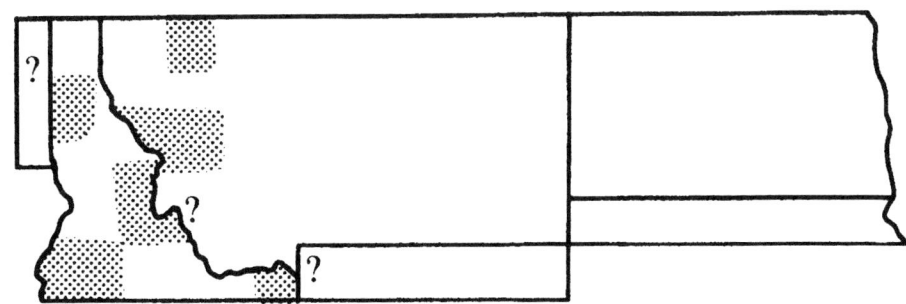

BURROWING OWL

Speotyto cunicularia
Strigidae

Summer Resident

WINTERING AREA: 4

HABITAT REQUIREMENTS: Prairie, grassland, meadow, open shrubsteppe, but does not use structurally similar montane habitats. Nests in abandoned mammal (usually ground squirrel or prairie dog) burrow. Also known to use habitat provided by airports and golf courses if relatively free of human disturbance.

FEEDING: Largely insectivorous but also frequently takes small mammals, which not uncommonly comprise majority of diet; also takes birds and other small vertebrates. Hunts from elevated perch or from low, patrolling flight, often hovering briefly before dropping onto prey; also stalks invertebrate prey on the ground. Often forages in the daytime.

STATUS AND MANAGEMENT: Numbers are small but relatively stable in Idaho and North Dakota, although populations continue to exhibit long-term declines in Montana. Elsewhere, numbers continue to decline steadily throughout the species range, although the data are equivocal in some parts of the West. Listed as a Threatened species in Canada. As a consequence of control measures that have reduced populations of colonial rodents and as prairie and plains habitats have been converted to agriculture, Burrowing Owls have been reduced greatly, as well. Favors well-grazed, early successional grasslands (typified by conditions found in prairie dog colonies) with soils having significant sand content (MacCracken et al., 1985). Management recommendations include maintenance of pesticide- and herbicide-free areas of 600 m radius around nest burrows and provision of uncultivated plots supporting dense herbaceous vegetation within owl home ranges to supply habitat for rodent and insect prey (Haug and Oliphant, 1990). Burrowing Owls will use artificial nest burrows (Collins and Landry, 1977).

FURTHER READING: Green and Anthony, 1989; Johnsgard, 1988; Konrad and Gilmer, 1984; Marti, 1974; Rich, 1986.

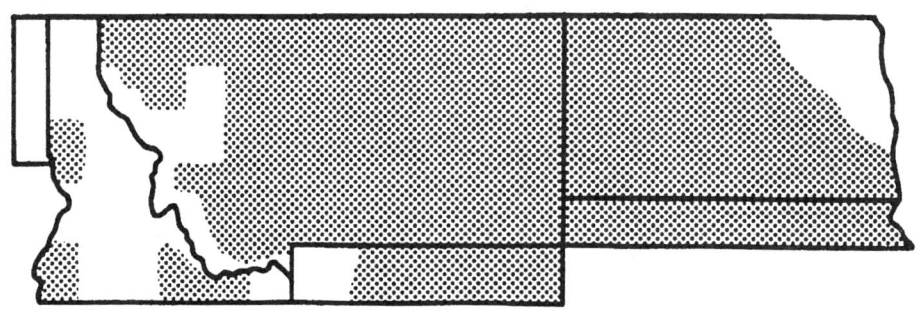

LONG-EARED OWL

Asio otus
Strigidae

Summer Resident but occasionally Permanent Resident in most of region (less commonly than Short-eared Owl)

WINTERING AREA: 5

HABITAT REQUIREMENTS: Coniferous or mixed forest, also frequently in aspen parkland or mixed woodland, dense thickets of small trees near open habitats, orchards, riparian woodland, and isolated woodlots; often near water. Nests in abandoned corvid or hawk nest or in mistletoe broom in trees.

FEEDING: A specialist on small mammals, primarily rodents (especially voles or, in some regions, pocket gophers or heteromyid rodents), only occasionally taking small birds, and rarely feeding on other small vertebrates or large insects. Drops onto prey from a low, searching flight, often systematically quartering over fields like a Short-eared Owl. Hunts primarily at night.

STATUS AND MANAGEMENT: Long-eared Owls are virtually absent from BBS surveys. Highly nomadic, local numbers fluctuate markedly in concert with abundance of primary rodent prey (especially responsive to vole cycles). Loss of riparian habitats in the West are likely to have resulted in decreased numbers in the region (Johnsgard, 1988). In northeastern Oregon, Long-eared Owls nest in extensive, dense, unlogged conifer forests and are dependent on dwarf-mistletoe brooms in Douglas fir for nest sites (Bull et al., 1989). In winter, often roosts communally in dense conifers or other suitable dense trees or shrubs, with favored sites used perennially.

FURTHER READING: Bosakowski et al., 1989; Craig et al., 1988; Marks, 1986; Marti, 1976.

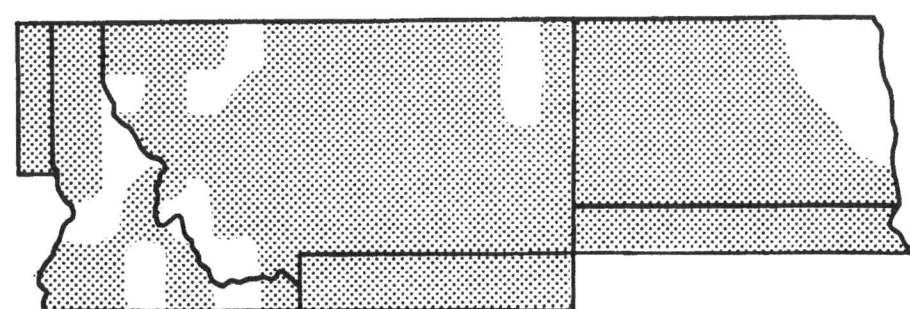

SHORT-EARED OWL

Asio flammeus
Strigidae

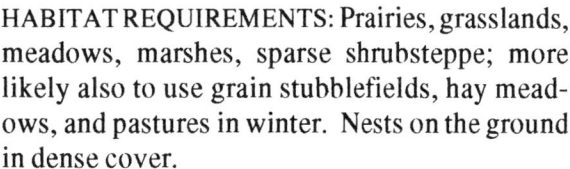

Permanent Resident in most of region but largely a Summer Resident in North Dakota

WINTERING AREA: 5

HABITAT REQUIREMENTS: Prairies, grasslands, meadows, marshes, sparse shrubsteppe; more likely also to use grain stubblefields, hay meadows, and pastures in winter. Nests on the ground in dense cover.

FEEDING: Almost exclusively rodents, especially voles, but also other small mammals and occasional birds and large insects. Forages in a prolonged, low, searching flight, often hovering briefly before dropping onto prey, or glides down onto prey from slightly elevated perch. Often seen hunting in the daytime.

STATUS AND MANAGEMENT: The nomadic behavior of Short-eared Owls creates difficulty for accurate assessments of its numbers. BBS data indicate only slight declines in Idaho and Montana and relative stability of low numbers in North Dakota, where it has declined markedly since the 1930s. This owl appears too infrequently to project population trends for its entire range, but the species has declined significantly in the West as favored marsh habitats have been destroyed and native prairies have been overgrazed or converted to agriculture; strongly associated with ungrazed and undisturbed native grasslands (Duebbert and Lokemoen, 1977; Kantrud and Higgins, 1992). Irruptive and nomadic, Short-eared Owls seek out areas with high rodent densities for breeding. Roosts singly or communally on the ground at all times of year, but switches to dense conifers for roosting in winter when snow precludes ground roosting.

FURTHER READING: Bosakowski, 1986; Clark, 1975; Holt et al., 1992; Johnsgard, 1988.

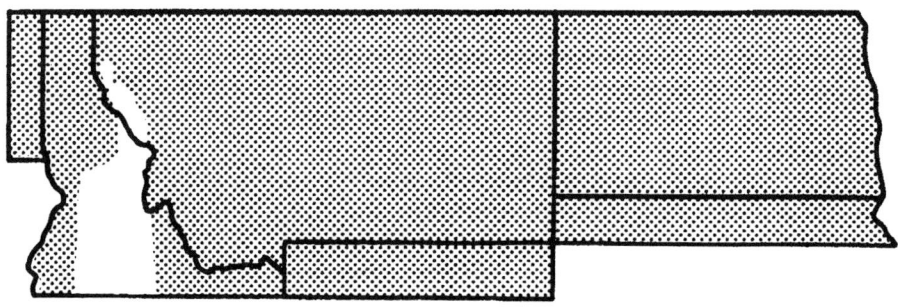

COMMON NIGHTHAWK

Chordeiles minor
Caprimulgidae

Summer Resident

WINTERING AREA: 1

HABITAT REQUIREMENTS: Open and partially open habitats: prairie, grassland, meadows, shrubsteppe, open woodland, agricultural fields, towns. Usually nests on the ground, occasionally on flat, graveled roof, and rarely on low stump or old passerine cup nest.

FEEDING: Forages aerially in continuous flight for a diet composed exclusively of flying insects. This is the only caprimulgid in North America that commonly forages in the daytime, although it is largely crepuscular (i.e., active at dawn and at dusk).

STATUS AND MANAGEMENT: Numbers have improved significantly in North Dakota and appear to be increasing slightly in Idaho but decreasing somewhat in Montana. For the species as a whole in North America, populations appear to be stable overall, although there have been areas of marked decreases in the most recent survey years, including the Dakotas, Wyoming, and Colorado, and a trend toward decreasing numbers in the West generally. Although flat-topped roofs are used both as roost sites and as nest sites, nighthawks strongly prefer natural sites, at least in some parts of their range (Brigham, 1989).

FURTHER READING: Aldridge and Brigham, 1991; Brigham, 1990.

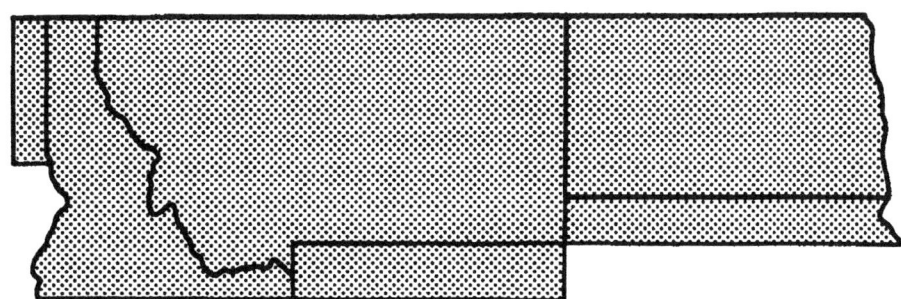

COMMON POORWILL

Phalaenoptilus nuttallii
Caprimulgidae

Summer Resident

WINTERING AREA: 5

HABITAT REQUIREMENTS: Arid or semiarid open, rocky habitats: prairie, badlands, canyons, open woodland, shrubsteppe, occasionally breeding in suitable montane habitat but only at lower elevations. Nests on the ground.

FEEDING: Forages in the air within a few meters of the ground by sallying from the ground to capture flying insects.

STATUS AND MANAGEMENT: Although distributed widely in the region, appears in BBS data only for Idaho, where it seems to be declining (but sample sizes are exceedingly small). BBS data are insufficient to assess overall trends from the species' range as a whole, but what few data there are indicate relative stability. Strictly nocturnal, foraging mostly at dusk and at dawn (i.e., a crepuscular forager), and relatively little known ecologically.

FURTHER READING: Brigham and Barclay, 1992.

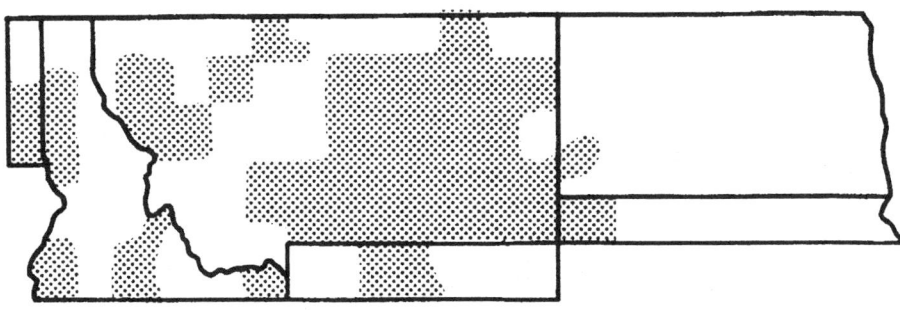

CHIMNEY SWIFT

Chaetura pelagica
Apodidae

Summer Resident

WINTERING AREA: 1

HABITAT REQUIREMENTS: Woodland and other open areas, especially near human dwellings, most often nesting in chimneys or other similarly shaped, human-built structures. Presumably nested in hollow trees prior to the arrival of European architecture in North America.

FEEDING: Dines on the wing, feeding exclusively on flying insects.

STATUS AND MANAGEMENT: Numbers are small but increasing significantly in North Dakota. Overall, populations appear to be declining in many areas, markedly so in the most recent survey years. Provision of suitable nesting and roosting sites in the form of chimneys likely has been responsible for a pronounced westward range extension by Chimney Swifts. Breeds in small colonies of a few pairs but gathers into enormous postbreeding flocks that roost together in suitable sites prior to migration.

FURTHER READING: Dexter, 1981; Zammuto et al., 1981.

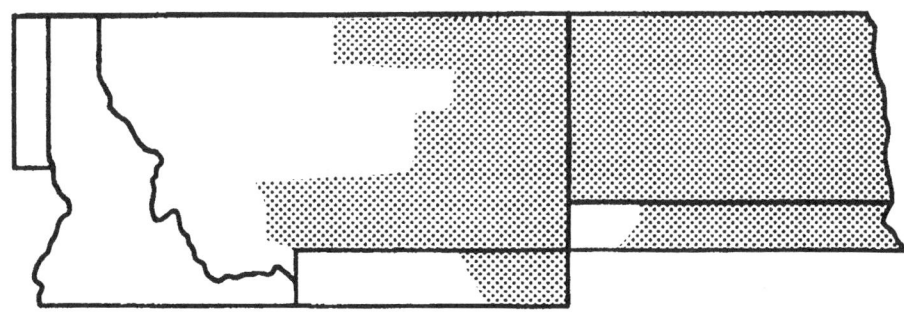

VAUX'S SWIFT

Chaetura vauxi
Apodidae

Apodidae

Summer Resident

WINTERING AREA: 3

HABITAT REQUIREMENTS: Coniferous forests, especially old-growth, but will nest in snags found in burned or logged forest, although these areas appear to be much less preferred. Nests in woodpecker-excavated cavity in heartwood fungus-infected live conifer (especially in cavities created by Pileated Woodpeckers in old-growth conifers) or in hollow coniferous snag; only seldom uses abandoned chimney for nest site. Reportedly nests in broken-top snags of older cottonwoods in western Montana cottonwood stands.

FEEDING: Forages aerially on a diet composed exclusively of flying insects.

STATUS AND MANAGEMENT: Appears infrequently in BBS database but numbers appear to be declining somewhat in both Idaho and Montana, as well as in the West as a whole. Vaux's Swifts appear to depend heavily on Pileated Woodpeckers for provision of suitable nest sites in old-growth coniferous forests (Bull and Cooper, 1991). Closely associated with old-growth Douglas fir/ponderosa pine forests providing large-diameter, broken-top snags (Mannan and Meslow, 1984). Roost sites are used communally from time of arrival on the breeding grounds until departure on autumn migration; suitable roost site is a large-diameter, hollowed snag or stub of old-growth conifer (Bull, 1991).

FURTHER READING: Bull and Hohmann, 1993.

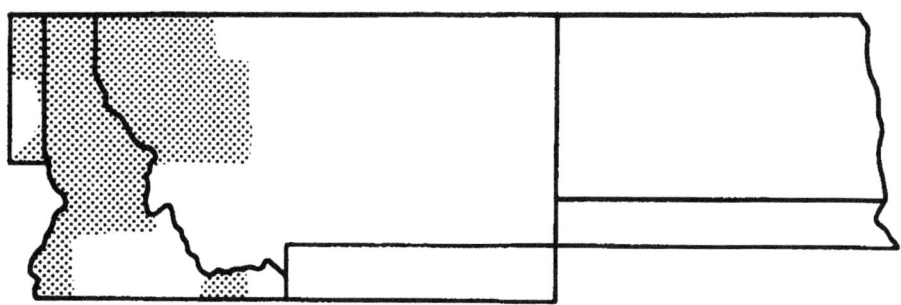

BLACK SWIFT

Cypseloides niger
Apodidae

Summer Resident

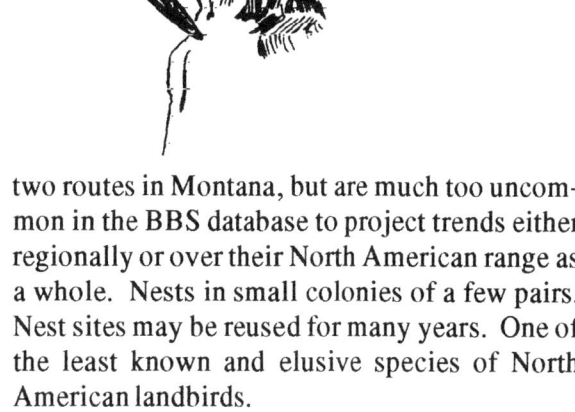

WINTERING AREA: 3

HABITAT REQUIREMENTS: Montane areas, associated with steep cliffs or canyons, especially near water. Nests in high rock crevice or shallow cave, not uncommonly behind a waterfall.

FEEDING: Feeds on the wing, dining exclusively on flying insects. Generally forages high in the air but will often fly low over rivers and streams in pursuit of emerging aquatic insects.

STATUS AND MANAGEMENT: Black Swifts have been detected on two BBS routes in Idaho and on two routes in Montana, but are much too uncommon in the BBS database to project trends either regionally or over their North American range as a whole. Nests in small colonies of a few pairs. Nest sites may be reused for many years. One of the least known and elusive species of North American landbirds.

FURTHER READING: Hunter and Baldwin, 1962; Knorr, 1961; Kondla, 1973.

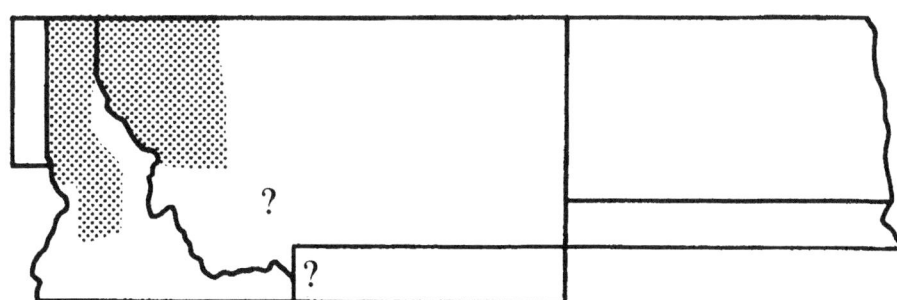

Apodidae

WHITE-THROATED SWIFT

Aeronautes saxatalis
Apodidae

Summer Resident

WINTERING AREA: 4

HABITAT REQUIREMENTS: Montane areas, nesting in rock crevices of sheer cliffs and canyon walls. Occasionally nests in tall, human-built structures, such as bell towers and grain elevators.

FEEDING: Feeds in flight, exclusively on flying insects. Generally forages high in the air over canyons or cliffs but will fly low over rivers and streams in pursuit of emerging aquatic insects.

STATUS AND MANAGEMENT: BBS data for the region are insufficient to make any assessment of trends, but in the West overall there has been a small but significant decline in numbers. Traditional use of nest and roost sites over many years has been documented (Dobkin et al., 1986). Postbreeding flocks roost communally prior to migration.

FURTHER READING: Collins, 1983.

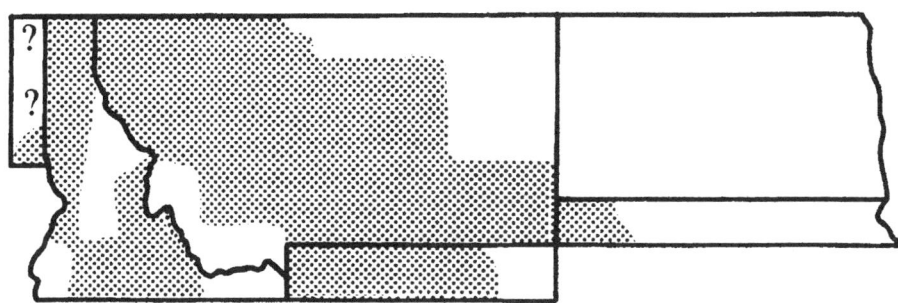

RUBY-THROATED HUMMINGBIRD

Archilochus colubris
Trochilidae

Summer Resident

WINTERING AREA: 4

HABITAT REQUIREMENTS: Brushy margins and openings in deciduous forest in mountains and on floodplains. Nests in trees near or over water, occasionally reusing nests in subsequent years.

FEEDING: Hovers to take floral nectar and glean insects and spiders from vegetation, but also relies at times on tree sap from sapsucker wells.

STATUS AND MANAGEMENT: Does not appear in the region's BBS database; North Dakota and northeastern Montana form the western edge of the Ruby-throated Hummingbird's breeding range in the U.S. Numbers have declined steadily throughout North America, which has been attributed to problems encountered on the tropical wintering grounds.

FURTHER READING: Johnsgard, 1983; Miller and Nero, 1983; Mulvihill et al., 1992; Willimont et al., 1988.

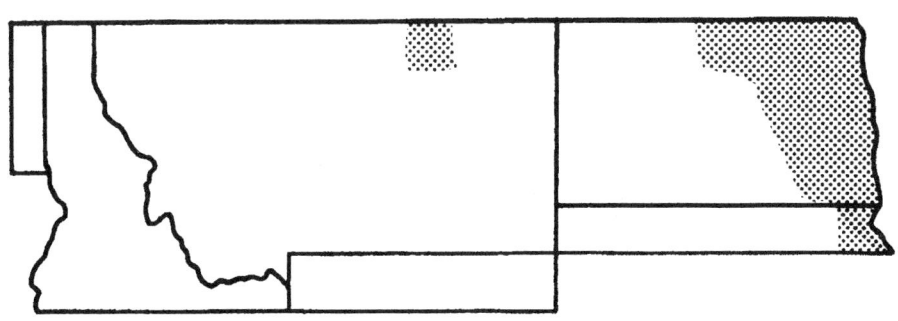

BLACK-CHINNED HUMMINGBIRD

Archilochus alexandri
Trochilidae

Trochilidae

Summer Resident

WINTERING AREA: 2

HABITAT REQUIREMENTS: Riparian woodland and open woodland near water in lower mountains, also montane meadows in migration. Generally nests in trees, usually near or over water.

FEEDING: Hovers to take floral nectar and glean insects and spiders from vegetation.

STATUS AND MANAGEMENT: Not very common in the region as reflected by the absence of BBS data. Over its entire range, numbers appear to be stable. Ecologically not well known in the region, which is the northern and eastern limit of the breeding range.

FURTHER READING: Baltosser, 1989; Ewald, 1985; Johnsgard, 1983; Stromberg and Johnsen, 1990.

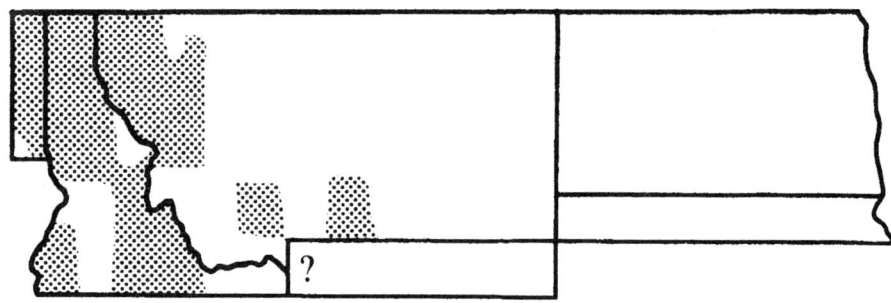

CALLIOPE HUMMINGBIRD

Stellula calliope
Trochilidae

Summer Resident

WINTERING AREA: 2

HABITAT REQUIREMENTS: Open montane forests, meadows, burned areas, and riparian thickets, also alpine tundra following breeding and in migration. Nests in trees, often building on old nests in successive years.

FEEDING: Hovers to take floral nectar and glean insects and spiders from vegetation.

STATUS AND MANAGEMENT: Numbers suggest stability of populations in the region, although sample sizes are very small. Overall, populations appear stable although numbers have decreased significantly in British Columbia in the most recent survey years.

FURTHER READING: Tamm et al., 1989; Tyrell and Tyrell, 1984.

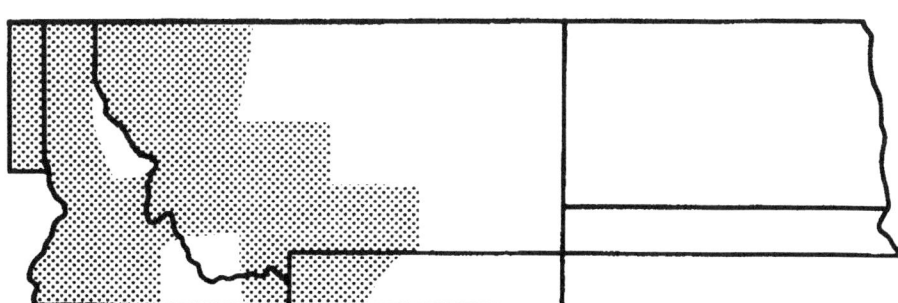

BROAD-TAILED HUMMINGBIRD

Trochilidae

Selasphorus platycercus
Trochilidae

Summer Resident

WINTERING AREA: 3

HABITAT REQUIREMENTS: Open woodland, brushy slopes, riparian and montane thickets, also other open habitats in migration. Generally nests in trees, often over or near mountain streams.

FEEDING: Hovers to take floral nectar and glean insects and spiders from vegetation.

STATUS AND MANAGEMENT: Numbers are too small for the region to give any indication of status, but appear relatively stable for the species range as a whole. The northern edge of the Broad-tailed Hummingbird's breeding range extends to the westernmost portion of the region.

FURTHER READING: Calder et al., 1983; Inouye et al., 1991; Johnsgard, 1983; Miller and Inouye, 1983; Waser, 1976.

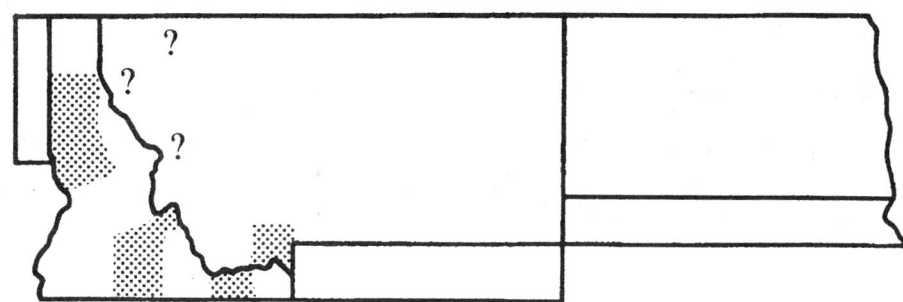

RUFOUS HUMMINGBIRD

Selasphorus rufus
Trochilidae

Summer Resident

WINTERING AREA: 2

HABITAT REQUIREMENTS: Open, montane coniferous forests, woodland edges, and thickets; in migration, nearly any open habitat providing abundant flowers for foraging. Nests usually in conifers but also uses deciduous trees or vine tangles; nests are often refurbished and reused in succeeding years.

FEEDING: Hovers to take floral nectar and glean insects and spiders from vegetation.

STATUS AND MANAGEMENT: BBS data are too sparse for the region to determine status, but numbers appear to be declining significantly in the West.

FURTHER READING: Calder and Jones, 1989; Carpenter and Hixon, 1988; Gass, 1979; Hiebert, 1991; Johnsgard, 1983; Paton and Carpenter, 1984; Tyrell and Tyrell, 1984.

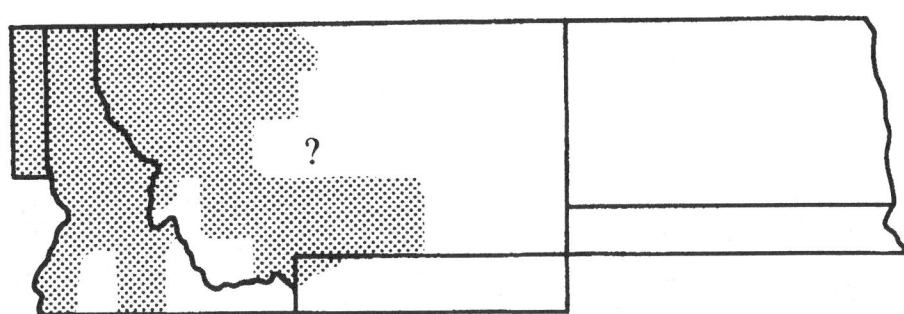

BELTED KINGFISHER

Ceryle alcyon
Alcedinidae

Primarily a Summer Resident but not uncommonly a Permanent Resident from central and southern Montana westward in the region

WINTERING AREA: 4

HABITAT REQUIREMENTS: From lowlands to montane parks, always associated closely with water, especially wooded streams and rivers, but also lakes, ponds, and any other permanent body of water providing fish. Nests in horizontal burrow excavated in vertical bank along streams and rivers. Prefers nest sites in soils with high sand and low clay composition. Usually excavate their own burrow or sometimes modify an abandoned mammal burrow; also known to rarely nest in cavity of a snag.

FEEDING: Diet is composed largely of small fish but also takes a wide variety of aquatic invertebrates, as well as occasional small vertebrates. Captures aquatic prey by diving into water from a perch or after hovering briefly while in flight; prey are captured with the beak.

STATUS AND MANAGEMENT: Populations in the region appear to be stable except in Idaho where there has been a small but significant continuing decline. Slight declines are indicated for Western populations as a whole, as well as for the species generally across the continent, although the picture is rather mixed.

FURTHER READING: Brooks and Davis, 1987; Davis, 1982, 1986.

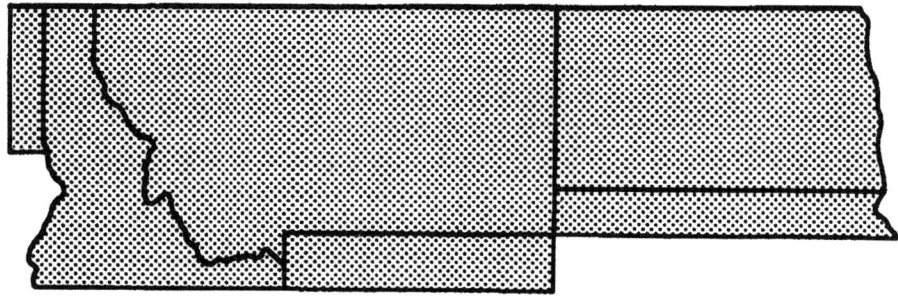

LEWIS' WOODPECKER

Melanerpes lewis
Picidae

Summer Resident, only rarely a Permanent Resident in western and southern Montana

WINTERING AREA: 5

HABITAT REQUIREMENTS: Open forest and woodland, primarily of ponderosa pine, including logged or burned areas with abundant snags, as well as riparian woodland (especially cottonwood groves). Nests in cavity of a dead limb on live tree or in standing snag, frequently used perennially.

FEEDING: Mainly insects caught flycatcher-like by sallying from a perch and capturing insects in flight, but also gleans food from the ground and from tree bark. Diet (especially in late summer and autumn) includes nuts, berries, pine seeds, fruit; caches acorns and nuts in natural bark crevices. Lewis' Woodpeckers rarely drill into bark to excavate insects like typical woodpeckers.

STATUS AND MANAGEMENT: An increasingly uncommon species in the region, as reflected by relatively infrequent encounters on the BBS routes; declining significantly in Montana and apparently declining in Idaho, as in the West generally. Logged or burned areas with standing snags provide suitable habitat only if a shrub understory is present. Populations using riparian woodlands in arid and semi-arid areas have declined precipitously in apparent response to loss and degradation of these riparian habitats by livestock; competition with European Starlings for suitable nest cavities may also be a factor.

FURTHER READING: Bock, 1970; Sousa, 1983.

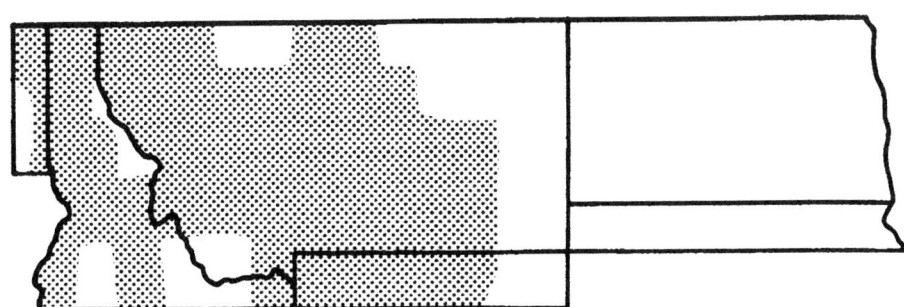

YELLOW-BELLIED SAPSUCKER

Sphyrapicus varius
Picidae

Summer Resident

WINTERING AREA: 4

HABITAT REQUIREMENTS: Extensive tracts of deciduous forests (especially riparian) in uplands (aspen, birch, poplar, ash) and lowlands (elm, ash, basswood, hackberry). Excavates cavity for nest in live deciduous tree, favoring trees infected with heart-rot fungus *(Fomes)*; does not reuse nest cavities but may excavate new cavity in favored tree in successive years.

FEEDING: Primarily feeds on insects, captured by bark gleaning and by drilling into trees; also consumes sap phloem tissue from deciduous trees (from conifers in early spring), and occasionally fruits and berries. Yellow-bellied Sapsuckers also infrequently will sally from a perch to hawk insects in flycatcher-like fashion.

STATUS AND MANAGEMENT: North-central and eastern North Dakota form the western edge of the Yellow-bellied Sapsucker's breeding range in the U.S. Data are insufficient from the region to assess status, but the species is declining significantly within its range as a whole, especially in the north-central portion of its distribution.

FURTHER READING: Kilham, 1971; Lawrence, 1967; Tate, 1973.

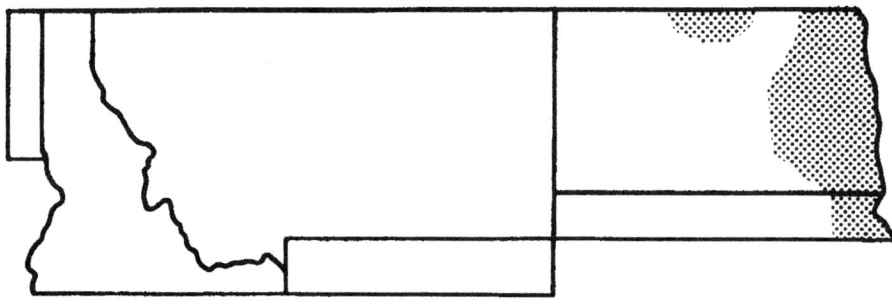

RED-NAPED SAPSUCKER

Sphyrapicus nuchalis
Picidae

Summer Resident

WINTERING AREA: 4

HABITAT REQUIREMENTS: Mature coniferous forest with aspen and montane riparian woodland, but most strongly associated with mature aspen (especially riparian) woodland. Excavates cavity for nest in live or dead deciduous tree (including larch, but only rarely in other coniferous trees), favoring aspens infected with heart-rot fungus (*Fomes*); does not reuse nest cavities but may excavate new cavity in favored tree in successive years.

FEEDING: Primarily feeds on insects, captured by bark gleaning and by drilling into trees; also consumes sap and pitch, occasionally fruits and berries. Relies heavily on sap from conifers as food source upon arrival in spring; sap of aspen and birch used only after buds open on the trees. Red-naped Sapsuckers also sally from perches to hawk insects in flycatcher-like fashion.

STATUS AND MANAGEMENT: BBS data are confounded by the recent division of the Yellow-bellied Sapsucker into two species, with the western populations now known as the Red-naped Sapsucker. Populations appear to be stable, but the heavy dependence of this species on mature riparian woodland in the more arid parts of its range may be cause for concern, as these habitats have been impacted heavily by livestock in many areas. In montane riparian habitats where it is the most abundant woodpecker, Red-naped Sapsuckers often function as "keystone" species by providing nest cavities for secondary cavity-nesting species (Dobkin and Wilcox, 1986). In the central Rockies, reduction of uncut aspen stands and conifer invasion of aspen woodland are likely to negatively impact sapsucker populations (Finch and Reynolds, 1988). In the northern Rockies, relative abundance and fledging success in small (<16 ha) logged units that retain snags and live deciduous trees do not differ from adjacent large tracts of unlogged forest (Tobalske, 1992).

FURTHER READING: Crockett and Hadow, 1975; Johnson and Zink, 1983; Mannan and Meslow, 1984.

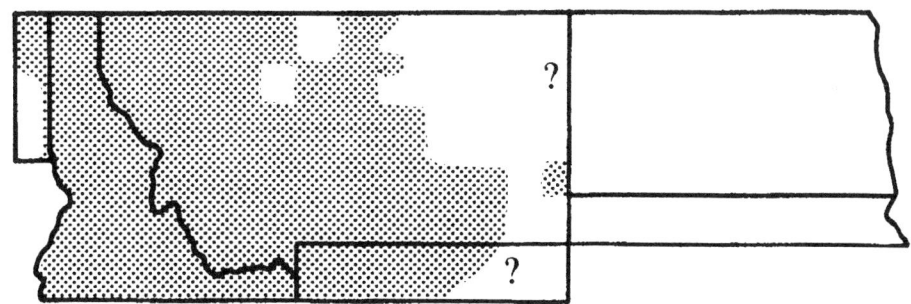

WILLIAMSON'S SAPSUCKER

Sphyrapicus thyroideus
Picidae

Summer Resident

WINTERING AREA: 5

HABITAT REQUIREMENTS: Montane coniferous forest, especially Douglas fir, ponderosa pine, and lodgepole pine, but also aspen woodland. Excavates cavity for nest in pine, aspen, fir, or larch; does not reuse nest cavities but may excavate new cavity in favored tree in successive years.

FEEDING: Primarily feeds on insects, especially ants, by drilling into trees; also consumes sap.

STATUS AND MANAGEMENT: Uncommon in the region and rarely encountered on the BBS routes, but what few data exist point to a significant decrease in numbers in Montana, the only state with data for the region. In the West as a whole, however, numbers appear to be increasing. Apparently an area-sensitive species, Williamson's Sapsucker is generally associated with old-growth forests (Mannan and Meslow, 1984; Moore, 1992) and is not found in smaller fragments (Aney, 1984).

FURTHER READING: Bock and Larson, 1986; Crockett and Hansley, 1977.

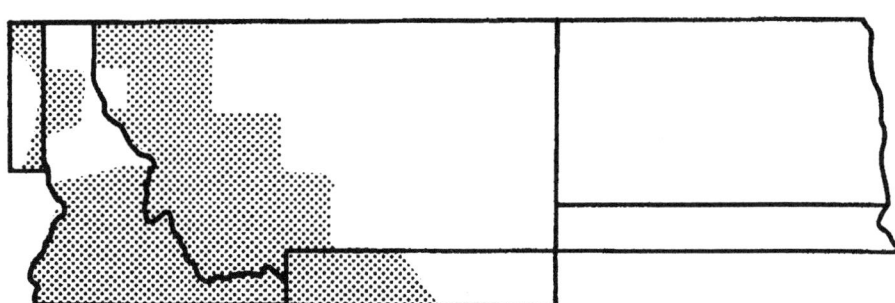

OLIVE-SIDED FLYCATCHER

Contopus borealis
Tyrannidae

Summer Resident

WINTERING AREA: 1

HABITAT REQUIREMENTS: Montane coniferous forest and woodland, especially burned areas with abundant standing snags, also riparian woodland. Usually nests in conifers.

FEEDING: Diet is composed exclusively of flying insects that are captured with sallies from exposed perches high in the tops of conifers.

STATUS AND MANAGEMENT: Numbers appear small but stable within the region, with a slight decline indicated in North Dakota, where it is at the edge of its breeding range and only infrequently encountered. In the West as a whole, however, Olive-sided Flycatchers have decreased significantly, as is also the case for its North American range in general. In forests of the northern Rockies, this flycatcher may depend heavily on early postfire habitats with an abundance of standing dead trees (Hutto, in prep.). Loss of wintering habitat in Central America has been implicated as an important contributor to the flycatcher's widespread decline.

FURTHER READING: Finch and Reynolds, 1988; Marshall, 1988.

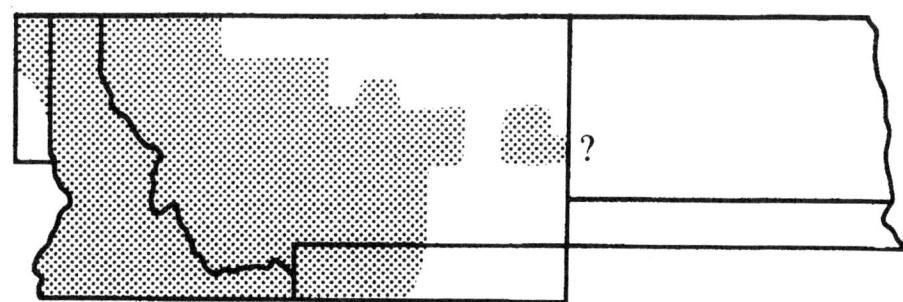

WESTERN WOOD-PEWEE

Contopus sordidulus
Tyrannidae

Summer Resident

WINTERING AREA: 1

HABITAT REQUIREMENTS: Coniferous and mixed coniferous-deciduous forest, woodland (especially riparian), and forest edge. Favors open forests dominated by conifers. Nests in conifer trees.

FEEDING: Sallies from exposed perch to capture insects in flight, then returns to perch, or hovers to glean insects or berries from vegetation.

STATUS AND MANAGEMENT: Numbers appear to be declining somewhat in Idaho but increasing significantly in Montana. The overall picture in the West is one of relative stability with perhaps a slight declining trend, although numbers have increased significantly in the most recent survey years. The foregoing picture stands in marked contrast to that of the very closely related and ecologically similar Eastern Wood-Pewee.

FURTHER READING: Beaver and Baldwin, 1975; Eckhardt, 1976; Hejl et al., 1988; Verbeek 1975a, b.

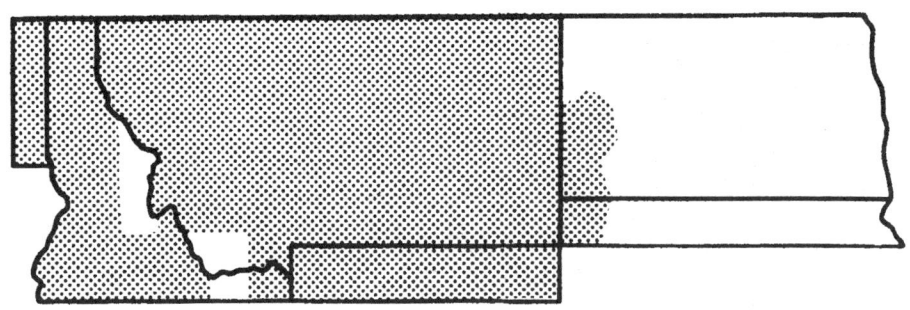

EASTERN WOOD-PEWEE

Contopus virens
Tyrannidae

Summer Resident

WINTERING AREA: 1

HABITAT REQUIREMENTS: Deciduous forest on floodplains and river bluffs along major watercourses and in extensive tracts of upland deciduous forest. Nests in deciduous trees.

FEEDING: Sallies from exposed perch to capture insects in flight, then returns to perch, or hovers to glean insects or berries from vegetation.

STATUS AND MANAGEMENT: No BBS data exist for Eastern Wood-Pewees in the region, but populations have decreased quite significantly across the species' North American range, especially in recent survey years.

FURTHER READING: Ehrlich et. al., 1988.

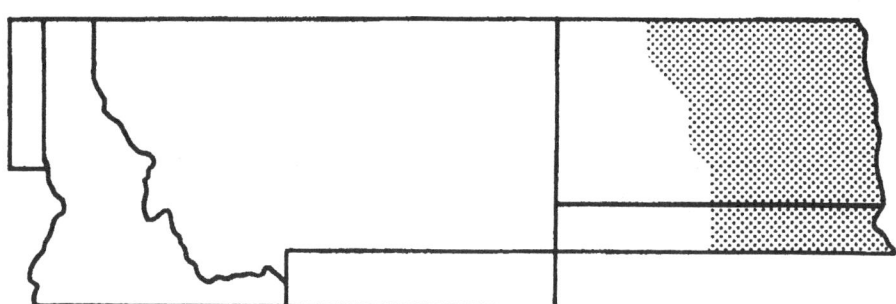

Tyrannidae

LEAST FLYCATCHER

Empidonax minimus
Tyrannidae

Summer Resident

WINTERING AREA: 3

HABITAT REQUIREMENTS: Open woodland and shrubland in mountains and lowlands, usually deciduous, especially floodplain forests in eastern portion of region, shelterbelts, parks, orchards. Nests usually in deciduous, sometimes coniferous, tree or shrub.

FEEDING: Feeds primarily by hovering to glean insects from vegetation and, less frequently, sallies from exposed perch to capture insects in flight, then returns to perch. Also hovers to glean berries from vegetation and occasionally includes a few seeds in diet.

STATUS AND MANAGEMENT: Numbers appear stable in North Dakota and significantly increasing in Montana. Least Flycatchers are found primarily to the east and north of the region, but appear to be expanding their range westward (N.K. Johnson, pers. comm.). The species is declining on a continent-wide basis (markedly so in recent survey years), but in contrast to the overall pattern, U.S. populations west of the Mississippi River appear to be increasing significantly. Least Flycatchers are sensitive to forest disturbances and respond either by 1) decreasing density of breeding birds in direct proportion to the intensity of logging operations (e.g., Freedman et al., 1981), or 2) shifting further into forest interior, retreating farther with increasing size of the disturbed area. (DellaSala and Rabe, 1987).

FURTHER READING: Briskie and Sealy, 1989; Darveau et al., 1992; Sherry and Holmes, 1988.

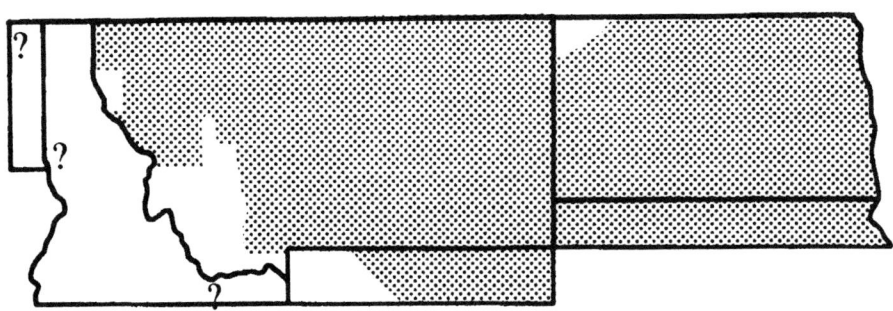

HAMMOND'S FLYCATCHER

Empidonax hammondii
Tyrannidae

Summer Resident

WINTERING AREA: 4

HABITAT REQUIREMENTS: Montane coniferous forest and woodland, favoring dense stands of fir or Douglas fir, also aspen and other deciduous, riparian woodland. Nests generally in live coniferous tree, occasionally in aspen or other deciduous tree.

FEEDING: Sallies from exposed, generally high, perch to capture insects in flight, then returns to perch. Diet composed exclusively of insects.

STATUS AND MANAGEMENT: Apparently declining in both Idaho and Montana but stable elsewhere in its range, although BBS data appear to fluctuate quite a bit. Hammond's Flycatchers are nesting-habitat specialists that strongly favor intact stands of old-growth (>200 yrs) and mature (99–199 yrs) coniferous forest, and are adversely affected by conversion of old-growth forests to younger age classes (Sakai and Noon, 1991). Although found in other age-classes of forest, the species is consistently more abundant in old growth (Hejl and Woods, 1991; Mannan and Meslow, 1984; Moore, 1992).

FURTHER READING: Beaver and Baldwin, 1975; Johnson and Marten, 1991; Manuwal, 1970.

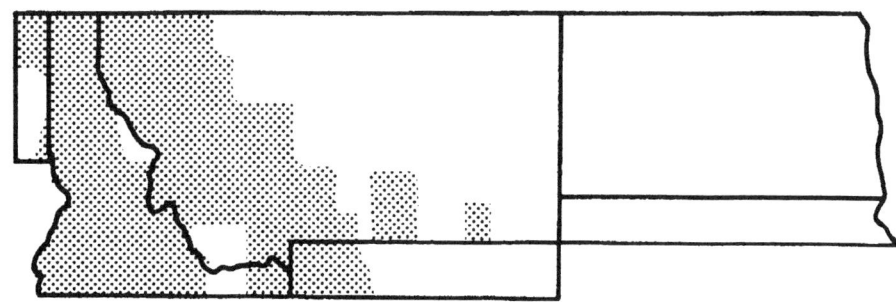

DUSKY FLYCATCHER

Empidonax oberholseri
Tyrannidae

Summer Resident

WINTERING AREA: 5

HABITAT REQUIREMENTS: Wide range of open woodland and shrub habitats: ponderosa pine or other conifers, mountain mahogany, aspen, riparian woodland, montane shrubsteppe. Nests most frequently in shrub, but also in low trees.

FEEDING: Sallies from exposed perch to capture insects in flight, then returns to perch, or hovers to glean insects from vegetation. Diet composed exclusively of insects.

STATUS AND MANAGEMENT: Numbers have increased significantly in Montana and appear stable in Idaho. Overall pattern throughout the range is one of stability with a trend toward marked increases in the most recent survey years. Consistently more abundant in managed coniferous forest stands than in old growth (e.g., Hejl and Woods, 1991; Mannan and Meslow, 1984). Often found on brushy, cutover slopes. Common cowbird host.

FURTHER READING: Johnson, 1966; Morton and Pereyra, 1985; Sedgwick, 1993.

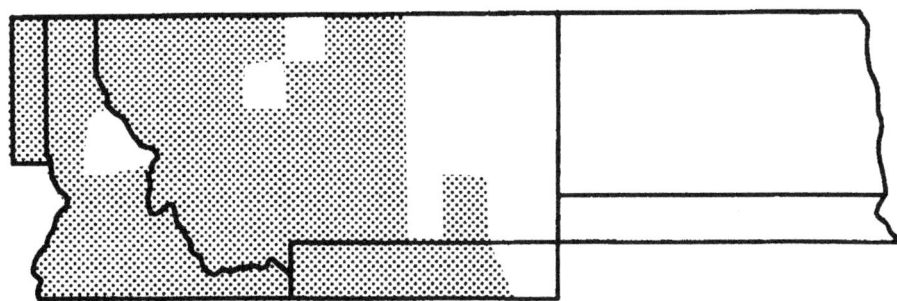

WILLOW FLYCATCHER

Empidonax traillii
Tyrannidae

Summer Resident

WINTERING AREA: 3

HABITAT REQUIREMENTS: Found at all elevations in dense willow thickets and other low, dense, riparian woodland, open woodland margins, and scattered shrub thickets or groves of trees (especially shelterbelts) within prairies. Nests in shrubs or low deciduous trees.

FEEDING: Sallies from exposed perch to capture insects in flight, then returns to perch, or hovers to glean insects or berries from vegetation; may occasionally include a few seeds in diet.

STATUS AND MANAGEMENT: Numbers appear stable in Idaho and increasing significantly in Montana and North Dakota. Numbers now appear to be increasing significantly for the species' range as a whole, with the exception of the Southwest and parts of the Midwest. Western populations have experienced a precipitous decline in recent years with southwestern populations approaching extinction. The southwestern population is listed as a candidate for Threatened or Endangered status at the federal level. Frequent cowbird host. Declines in the West have been attributed to a combination of riparian habitat degradation by livestock (Bock et al., in press) and heavy parasitism by cowbirds; may also be threatened by deforestation on the wintering grounds in the tropics. Populations increase in response to reduction of cattle grazing and cessation of willow control in riparian habitats (Taylor and Littlefield, 1986).

FURTHER READING: Barlow and McGillivray, 1983; Frakes and Johnson, 1982; Sanders and Flett, 1989; Sedgwick and Knopf, 1989.

Tyrannidae

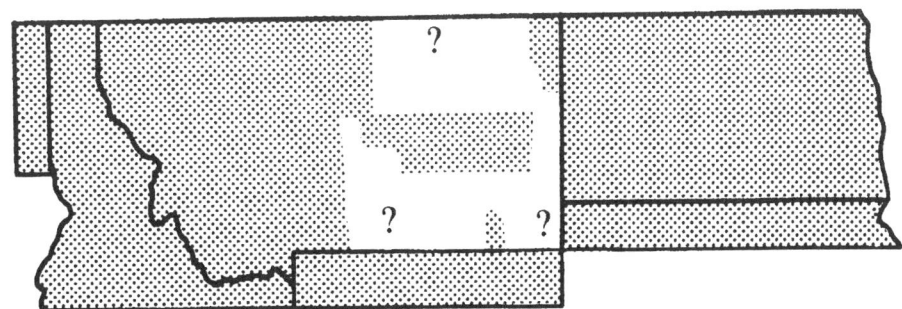

91

CORDILLERAN FLYCATCHER

Empidonax occidentalis
Tyrannidae

Summer Resident

WINTERING AREA: 2

HABITAT REQUIREMENTS: Found mainly in association with streams in montane coniferous forest, dense second-growth, aspen, and riparian woodlands, especially in canyons. Often nests along streams, usually in cavity of small tree or on cliff ledge, also on human-built structures and on or near the ground among tangled tree roots.

FEEDING: Sallies from exposed perch to capture insects in flight, then returns to perch, or hovers to glean insects or berries from vegetation; may occasionally include a few seeds in diet.

STATUS AND MANAGEMENT: Populations in Idaho have declined sharply and Montana numbers show a slight downward trend, as well. For the Cordilleran Flycatcher's range taken as a whole, numbers appear to be relatively stable.

FURTHER READING: Beaver and Baldwin, 1975; Johnson, 1980.

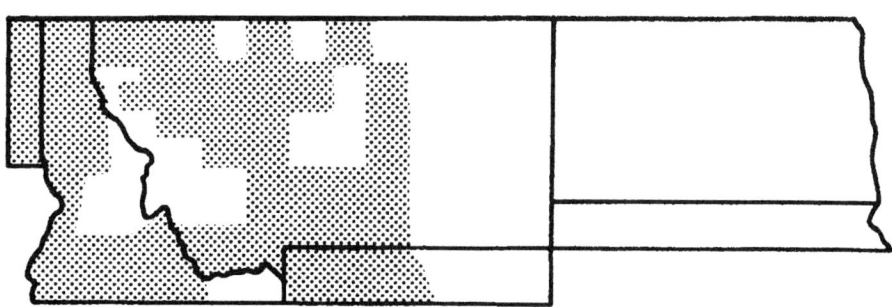

EASTERN PHOEBE

Sayornis phoebe
Tyrannidae

Summer Resident

WINTERING AREA: 5

HABITAT REQUIREMENTS: Open riparian or other woodlands adjoining streams and lakes, usually near bridges or abandoned buildings on which it often builds its nest.

FEEDING: Sallies from exposed perch to capture insects in flight (often low over water), then returns to perch; also reported to occasionally take small aquatic vertebrates.

STATUS AND MANAGEMENT: No BBS data exist for Eastern Phoebes in the region. Over the entire range there has been a steady, slight decline although numbers throughout the East were up significantly in the most recent survey year. A common cowbird host which, in combination with problems of habitat loss on the wintering grounds, may be responsible for recent declines.

FURTHER READING: Weeks, 1978.

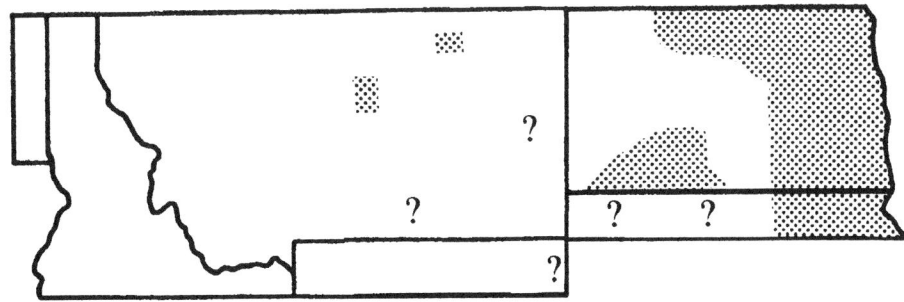

SAY'S PHOEBE

Sayornis saya
Tyrannidae

Summer Resident

WINTERING AREA: 5

HABITAT REQUIREMENTS: Arid and semiarid, open habitats including badlands, shrubsteppe, grasslands, open woodland, and agricultural lands, rarely in montane areas. Nest is built on cliff ledge, in abandoned building, or beneath eaves on buildings, under bridges, or in road culverts.

FEEDING: Sallies from exposed perch to capture insects in flight, then returns to perch, or hovers to glean insects or (rarely) berries from vegetation. Not uncommonly forages low over streams or ponds but frequently found well away from water.

STATUS AND MANAGEMENT: Numbers appear stable in most of the region. In Idaho, based on a rather small sample size, there appears to be a slight declining trend. In the West as a whole, the picture is mixed with a slight declining trend overall and sharper declines in the most recent survey years. Breeding biology and ecology are little known.

FURTHER READING: Ehrlich et al., 1988.

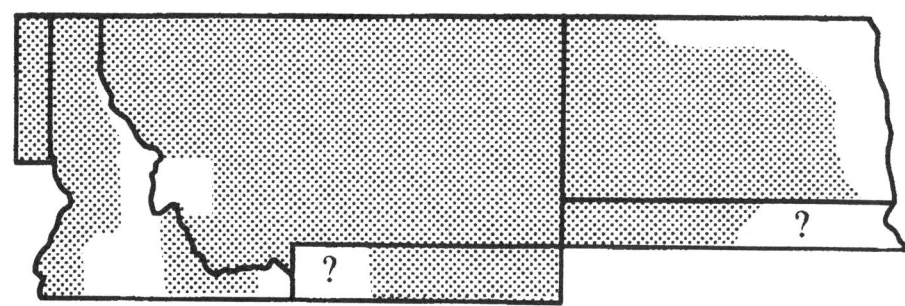

GREAT CRESTED FLYCATCHER

Myiarchus crinitus
Tyrannidae

Summer Resident

WINTERING AREA: 3

HABITAT REQUIREMENTS: Extensive tracts of mature floodplain, swamp, and upland deciduous forest with partially open canopies. Nests in natural or woodpecker-excavated cavities in live or dead deciduous trees.

FEEDING: Sallies from exposed perch to capture insects in flight, then returns to perch, or flies down to ground to capture insects, other terrestrial invertebrates, and small lizards. Includes small fruits in diet, especially berries, taken by hovering and picking them from plants.

STATUS AND MANAGEMENT: BBS data are scant for the region but numbers appear to be stable in North Dakota, as elsewhere overall in the North American range. Numbers appear to have decreased significantly in the Northeast and increased significantly in the upper Midwest in the most recent survey years. Great Crested Flycatchers display relatively narrow habitat affinities for mature open deciduous forest, which in combination with their dependence on cavities for nesting, make them a species deserving of close monitoring. Highly vulnerable to tropical deforestation (Morton, 1992).

FURTHER READING: Ehrlich et al., 1988.

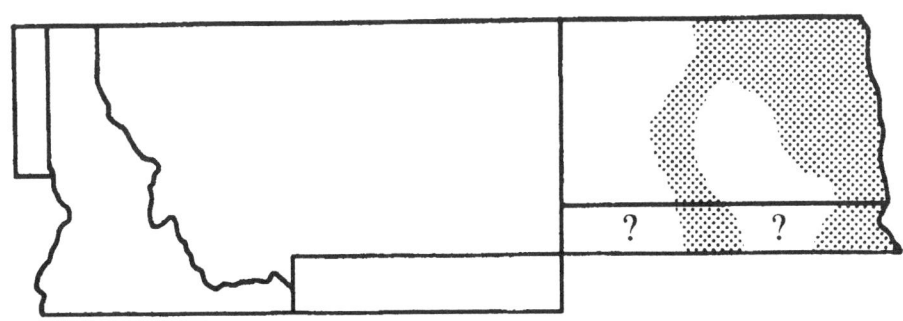

CASSIN'S KINGBIRD

Tyrannus vociferans
Tyrannidae

Summer Resident

WINTERING AREA: 3

HABITAT REQUIREMENTS: Dry savannah, open scrub, shrubsteppe, and especially pinyon/juniper woodland. Nests in trees.

FEEDING: Sallies from exposed perch to capture insects in flight, then returns to perch, also hovers and pounces on terrestrial insects. Includes a relatively large amount of small fruits in diet, often taken by hovering and picking them from plants.

STATUS AND MANAGEMENT: Southeastern Montana is the extreme northern limit of this species' breeding range. No BBS data exist for Cassin's Kingbird in the region; appears to be declining slightly in the West but sample sizes are relatively small.

FURTHER READING: Blancher and Robertson, 1984, 1987; Ohlendorf, 1974.

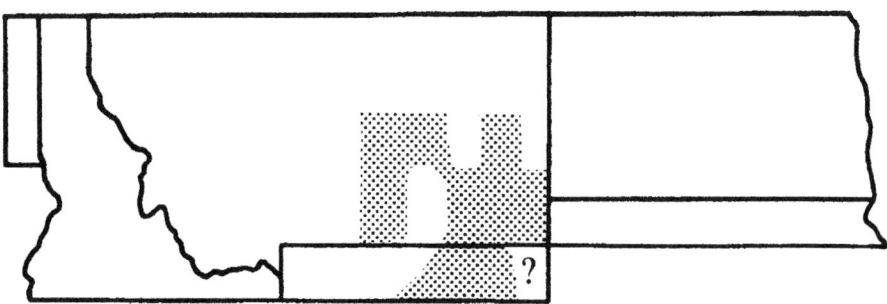

WESTERN KINGBIRD

Tyrannus verticalis
Tyrannidae

Summer Resident

WINTERING AREA: 3

HABITAT REQUIREMENTS: Open country, especially savannah, forest edge, shelterbelts, wooded residential areas, and agricultural lands with scattered trees and shrubs, also riparian woodland. Nests in trees (usually deciduous) and sometimes in shrubs or on human-built structures.

FEEDING: Sallies from exposed perch to capture insects in flight, then returns to perch, also hovers and pounces on terrestrial insects. Includes small fruits in diet, especially berries, taken by hovering and picking them from plants.

STATUS AND MANAGEMENT: Numbers are stable or increasing slightly, significantly so in North Dakota, and exhibit a significant overall increase in the West viewed as a whole. Unlike most Neotropical migrant songbirds, Western Kingbirds have benefited greatly from the creation of edge-type habitats and have greatly expanded their range in the region since the early 1800s as a result of human-induced alterations of the landscape.

FURTHER READING: Blancher and Robertson, 1987; MacKenzie and Sealy, 1981; Murphy, 1988.

Tyrannidae

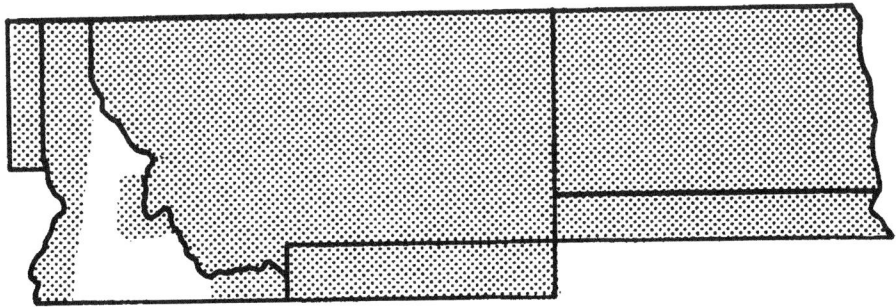

EASTERN KINGBIRD

Tyrannus tyrannus
Tyrannidae

Summer Resident

WINTERING AREA: 1

HABITAT REQUIREMENTS: Forest edge and open areas with scattered trees and shrubs, riparian woodland, agricultural lands with bushes and fence rows, shelterbelts, parks and wooded residential areas. Nests in deciduous tree, often isolated, and occasionally in shrub, on fence post, stump, or human-built structure.

FEEDING: Sallies from exposed perch to capture insects in flight, then returns to perch; also hovers and gleans insects from vegetation. Includes small fruits in diet, often taken by hovering and picking them from plants.

STATUS AND MANAGEMENT: Numbers appear to be declining in Idaho (especially in the most recent survey years) and Montana but increasing significantly in North Dakota. In general, numbers are down in the West and down across the North American range in general, with marked recent decreases in the Northeast. Eastern Kingbirds are common hosts of cowbirds.

FURTHER READING: Hayes and Robertson, 1989; McKitrick, 1990; Murphy, 1988.

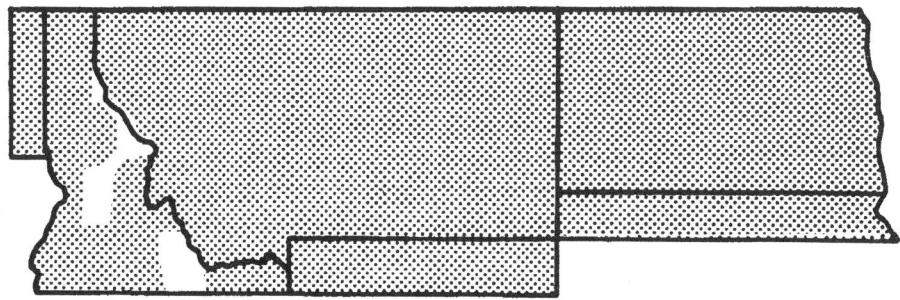

HORNED LARK

Eremophila alpestris
Alaudidae

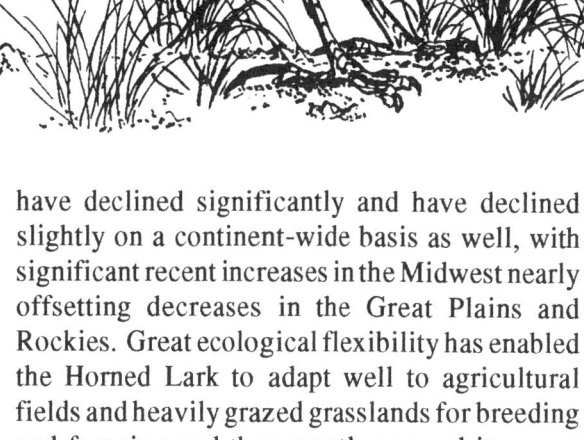

Permanent Resident throughout region except at highest elevations where it is a Summer Resident

WINTERING AREA: 6

HABITAT REQUIREMENTS: Open, treeless habitats from prairie to alpine tundra, including agricultural fields. Nests on the ground in a slight depression, with little or no cover.

FEEDING: Forages on the ground, gleaning seeds, insects, spiders, and snails.

STATUS AND MANAGEMENT: Numbers have decreased significantly in Idaho, increased significantly in Montana, and remained stable in North Dakota. For the West as a whole, numbers have declined significantly and have declined slightly on a continent-wide basis as well, with significant recent increases in the Midwest nearly offsetting decreases in the Great Plains and Rockies. Great ecological flexibility has enabled the Horned Lark to adapt well to agricultural fields and heavily grazed grasslands for breeding and foraging and thus greatly expand its range eastward since the early 1800s.

FURTHER READING: George et al., 1992; Hurley and Franks, 1976; Wiens et al., 1986.

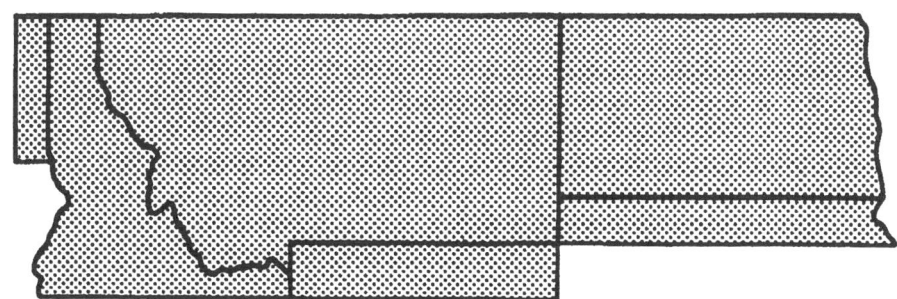

PURPLE MARTIN

Progne subis
Hirundinidae

Summer Resident from central North Dakota eastward, Migrant throughout region

WINTERING AREA: 1

HABITAT REQUIREMENTS: Open country near water, but limited in the region largely to areas where communal martin nest houses are provided, thus found mostly in the vicinity of human habitations. Until the early 1900s, nests generally were placed in tree cavities in standing snags or live deciduous trees, in cliff niches, or in other cavities.

FEEDING: Forages aerially in prolonged flight, capturing flying insects. Occasionally lands on the ground to feed on ants or other insects.

STATUS AND MANAGEMENT: Numbers appear to be declining in North Dakota; populations have declined markedly in the western portion of the U.S. but have stabilized for North America as a whole, although there are scattered significant declines in the most recent survey years. Has responded readily to provision of communal nest boxes for breeding, especially in the eastern part of range. The Purple Martin Conservation Association is currently conducting an intensive continent-wide effort to locate and monitor all breeding colonies.

FURTHER READING: Morton and Derrickson, 1990; Morton et al., 1990; Stutchbury, 1991.

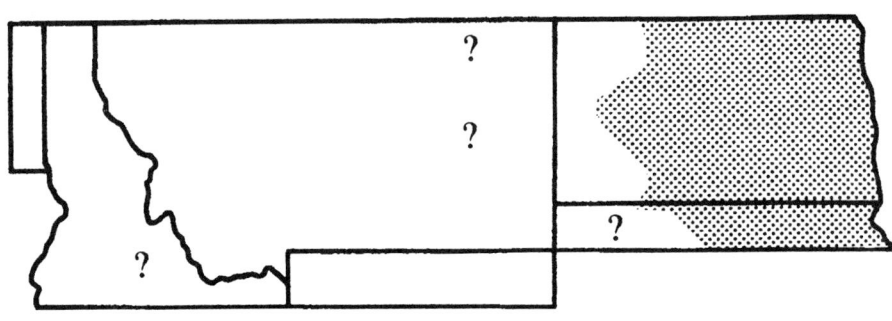

TREE SWALLOW

Tachycineta bicolor
Hirundinidae

Summer Resident

WINTERING AREA: 4

HABITAT REQUIREMENTS: Woodland of all types near water, especially with abundant standing snags. Nests in cavity in snag, also will use cavity in fence post or building.

FEEDING: Forages aerially in prolonged flight, capturing flying insects. When flying insects are unavailable, will glean berries or insects from vegetation.

STATUS AND MANAGEMENT: A trend of increasing numbers appears throughout the region, significantly so in the West as a whole, as well as throughout the North American range, although there have been significant declines in scattered parts of the range (including Idaho) in the most recent survey years. Favors riparian aspens for breeding sites in the region. As in other cavity-nesting species that cannot excavate their own nest sites (i.e., "secondary" cavity nesters), breeding population size can be limited by both intra- and interspecific competition for suitable nest cavities, as well as by differential susceptibility of cavities to predation, by snag fall, and by the relative abundance of woodpeckers (i.e., "primary" cavity nesters) capable of excavating new cavities (Rendell and Robertson, 1989). Often nests colonially and will readily use nest boxes for breeding. Nest site selection, especially in terms of proximity to forest edge, can be influenced strongly by presence of House Wrens, which favor nest sites close to forest edge and readily destroy Tree Swallow clutches (Rendell and Robertson, 1990). Forms large premigratory flocks that roost communally.

FURTHER READING: Lombardo, 1991; St. Louis and Breebaart, 1991; Wheelwright and Dorsey, 1991; Williams, 1988; Winkler, 1992.

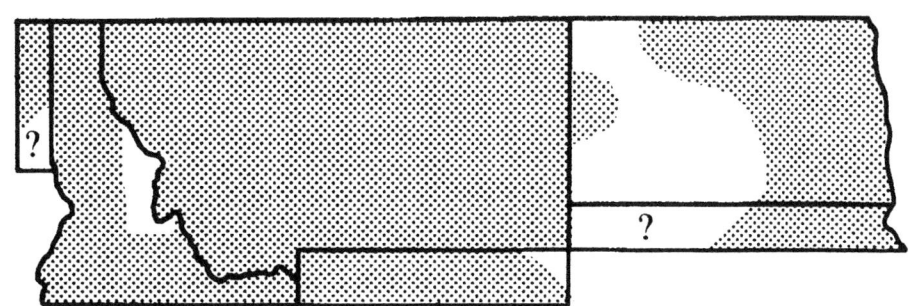

VIOLET-GREEN SWALLOW

Tachycineta thalassina
Hirundinidae

Summer Resident

WINTERING AREA: 4

HABITAT REQUIREMENTS: Open, montane coniferous forest, aspen and riparian woodland. Nests in natural or woodpecker-excavated cavities in standing snags or live trees, less frequently in cliff niche or abandoned building.

FEEDING: Forages aerially in prolonged flight, capturing flying insects. Only rarely comes to the ground to feed on insects.

STATUS AND MANAGEMENT: Increasing significantly both in Idaho and Montana, with a trend toward slight increase throughout its range in the west, significantly so in the most recent survey years. Occasionally nests in small colonies. Accepts nestbox as breeding site, especially in habitats where natural or woodpecker-excavated cavities are scarce or where interspecific competition for nest sites among secondary cavity nesters is high.

FURTHER READING: Brawn, 1990; Erskine, 1984.

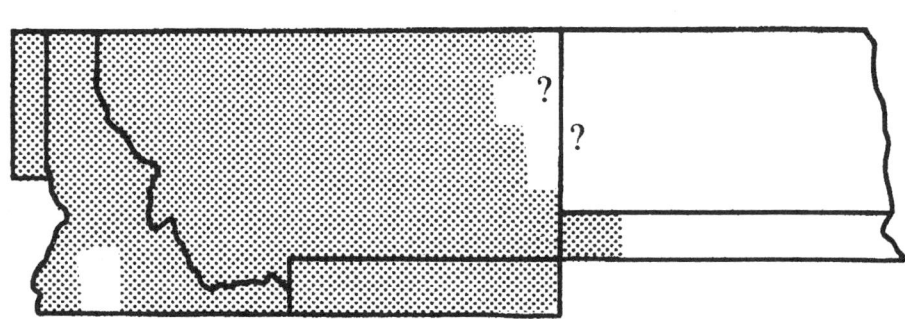

NORTHERN ROUGH-WINGED SWALLOW

Stelgidopteryx serripennis
Hirundinidae

Summer Resident

WINTERING AREA: 4

HABITAT REQUIREMENTS: Open areas in the vicinity of water, especially those having steep embankments or cliffs. Nests in burrows or crevices in steep river- or streambank, also in cliff niche, and in road culverts or beneath bridges. Nest burrows may be excavated by the birds but more often are abandoned burrows of mammals or kingfishers.

FEEDING: Forages aerially in prolonged flight, capturing flying insects. Occasionally descends to the ground to feed on insects.

STATUS AND MANAGEMENT: Numbers appear stable in Idaho and North Dakota but have declined significantly and rather sharply in Montana, but the general pattern in the West as a whole is one of stability, as is also the pattern continent-wide. However, significant declines have occurred in the most recent survey years in scattered parts of the range, including North Dakota. Nests in solitary pairs rather than colonially, although Rough-wings occasionally nest within Bank Swallow colonies.

FURTHER READING: Lunk, 1962; Ricklefs, 1972.

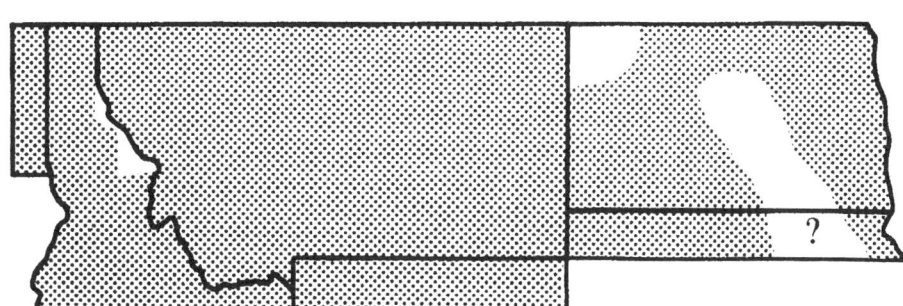

BANK SWALLOW

Riparia riparia
Hirundinidae

Summer Resident

WINTERING AREA: 1

HABITAT REQUIREMENTS: Open country in association with water, especially rivers and streams. Nests in burrows in steep river- or streambank, roadcut, or gravel pit. Pairs usually excavate their own nest burrow, but sometimes will use abandoned kingfisher burrow.

FEEDING: Forages aerially in prolonged flight, capturing flying insects. Only rarely lands on the ground to feed on insects.

STATUS AND MANAGEMENT: Regional status is very mixed: populations are increasing significantly in Idaho, decreasing significantly in Montana, and stable in North Dakota. Overall, western populations appear relatively stable as is also the case continent-wide for this widely distributed species. However, the most recent survey years exhibit steep, significant overall declines. Nests in colonies and forms large postbreeding flocks prior to migration, which roost communally.

FURTHER READING: Birchard and Kilgore, 1980; Freer, 1979; Garrison et al., 1987, 1989; Stutchbury, 1988.

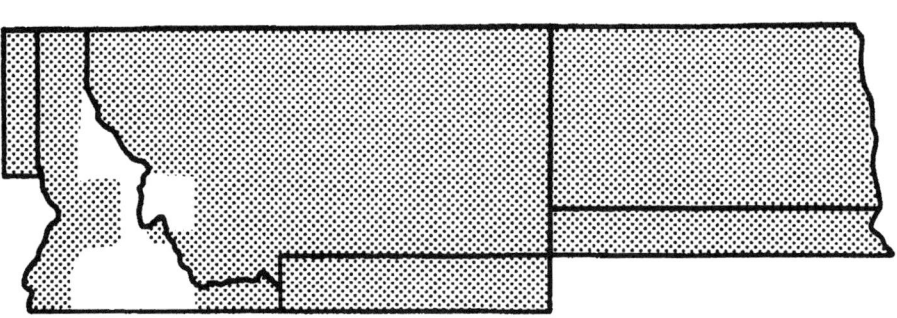

CLIFF SWALLOW

Hirundo pyrrhonota
Hirundinidae

Summer Resident

WINTERING AREA: 1

HABITAT REQUIREMENTS: Open country, usually in the vicinity of water. Nests on cliffs beneath overhangs, beneath bridges, in road culverts, and on buildings. Nests are often repaired and reused in subsequent years.

FEEDING: Forages aerially in prolonged flight, capturing flying insects. Occasionally feeds on berries.

STATUS AND MANAGEMENT: Slight increases are apparent in Idaho and Montana with significant increases in North Dakota. Western populations in general show a slight increasing trend. A few scattered areas in eastern North America show declines, some of which are significant in recent survey years, but the overall picture is relatively stable. Breeds colonially and may alternate colony site between years to reduce buildup of nest parasite populations.

FURTHER READING: Beecher et al., 1985; Brown and Brown, 1986, 1988; Shields, 1990; Withers, 1977.

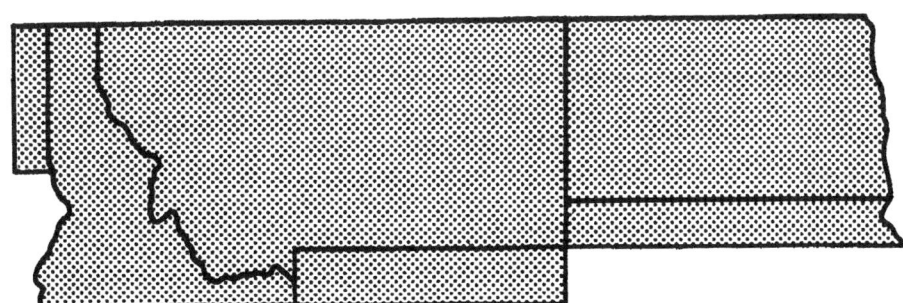

BARN SWALLOW

Hirundo rustica
Hirundinidae

Summer Resident

WINTERING AREA: 1

HABITAT REQUIREMENTS: Open areas, especially near water and frequently in association with human habitation and agricultural areas. Nests most often on or in buildings, also beneath bridges, in road culverts, and beneath overhangs on cliffs.

FEEDING: Forages aerially in prolonged flight, capturing flying insects. Occasionally feeds on berries and seeds.

STATUS AND MANAGEMENT: Numbers appear stable in Idaho and Montana but increasing significantly in North Dakota; Idaho populations show a significant increase in numbers in the most recent survey year. Overall in the West, numbers are stable and show a small but significant increase continent-wide, although there has been a significant widespread decline in the most recent survey years, which is most pronounced from Colorado eastward to the Middle Atlantic states and northward to southeastern Canada. Nesting is sometimes loosely colonial.

FURTHER READING: Grzybowski, 1979; Medvin et al., 1987; Shields, 1984; Snapp, 1976.

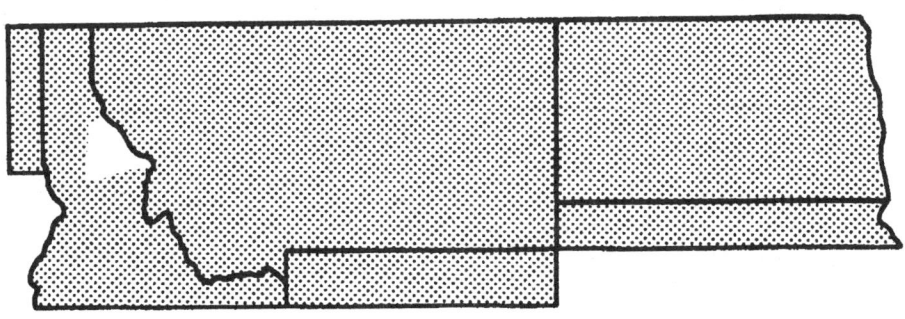

BROWN CREEPER

Certhia americana
Certhiidae

Permanent Resident from central Montana westward and occasional Winter Resident in North Dakota and eastern Montana

WINTERING AREA: 6

HABITAT REQUIREMENTS: Montane pine forests (especially ponderosa), mixed cedar/hemlock and other coniferous forest, and mixed coniferous/deciduous forests. Shuns pure or nearly-pure deciduous forests in central Rockies (Scott and Crouch, 1988b). Nests beneath loose bark on conifer tree, less frequently on deciduous tree.

FEEDING: Gleans insects, spiders, and other invertebrates from bark of tree trunks and branches, opportunistically hawks flying insects. Prefers larger trees as foraging substrates. Also infrequently takes nuts, berries, and acorns.

STATUS AND MANAGEMENT: Significantly declining in Idaho but status is unclear in Montana due to infrequency on BBS routes there. Western populations in general appear to be increasing slightly, as is also the case for continent-wide pattern. Appears sensitive to forest fragmentation during breeding season and may be considered as a forest-interior nesting species (Aney, 1984; Keller and Anderson, 1992; Mannan and Meslow, 1984).

FURTHER READING: Davis, 1978; Franzreb, 1985; Hejl et al., 1988.

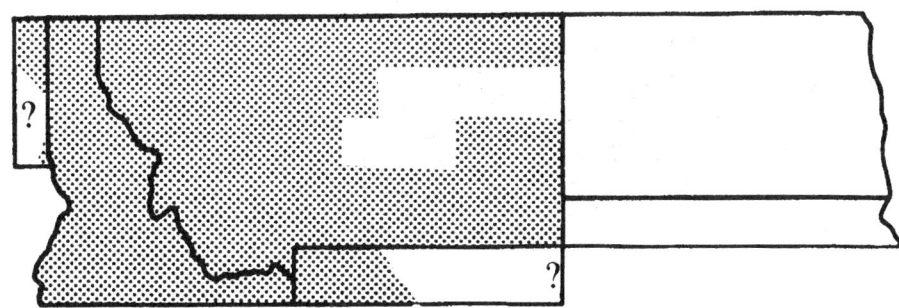

Troglodytidae

ROCK WREN

Salpinctes obsoletus
Troglodytidae

Summer Resident, rare Permanent Resident in westernmost portion of region

WINTERING AREA: 5

HABITAT REQUIREMENTS: Rocky habitats with brush or shrubs, especially cliff walls, talus slopes, and rock outcrops in montane areas, canyons, and badlands. Nests usually in rock crevices on steep slopes.

FEEDING: Forages for insects and other terrestrial invertebrates (possibly including small lizards) by gleaning items from the ground or low vegetation.

STATUS AND MANAGEMENT: Numbers appear to be increasing throughout the region, significantly so in Montana. In the West overall, however, there is a slight declining trend. Occasional cowbird host. Rock Wrens have been little studied in the field, resulting in many gaps in our knowledge of their ecology.

FURTHER READING: Tramontano, 1975; Wolf et al., 1985.

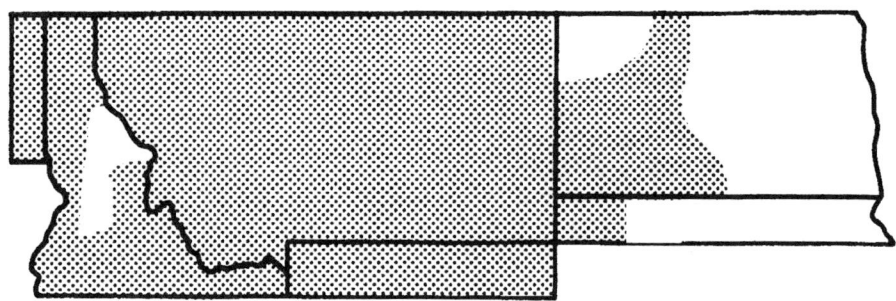

HOUSE WREN

Troglodytes aedon
Troglodytidae

Summer Resident

WINTERING AREA: 5

HABITAT REQUIREMENTS: Generally at lower elevations; riparian, aspen, and other sorts of open woodlands, also shrubby thickets, shelterbelts, farmlands, and areas of human habitation that are at least partially wooded. Nests in natural or woodpecker-excavated cavity in live tree, stump, or standing snag, but also will use a wide variety of other types of cavities.

FEEDING: Forages on the ground or in vegetation, gleaning insects and other terrestrial invertebrates from the surfaces of plants or directly from the ground.

STATUS AND MANAGEMENT: Populations appear stable in Idaho and Montana and significantly increasing in North Dakota. Western populations as a whole are increasing significantly, as is also the case when considering the species continent-wide. House Wrens are often the most abundant avian species in deciduous woodlands of the Rockies (Finch, 1989a, b). Populations respond positively to retention of logging slash, which supplies cover, foraging areas, and nest sites (Tobalske et al., 1991). This species often destroys the eggs of other songbirds (including conspecifics) in the vicinity of their nests. Readily uses nestboxes.

FURTHER READING: Drilling and Thompson, 1991; Finch, 1990, 1991a; Guinan and Sealy, 1989; Kermott et al., 1991; Rendell and Robertson, 1990.

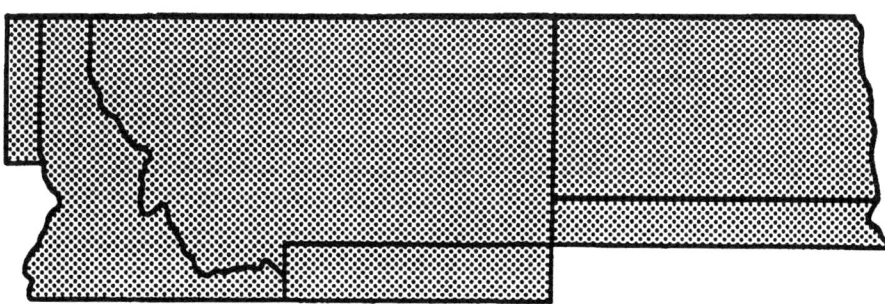

Troglodytidae

SEDGE WREN

Cistothorus platensis
Troglodytidae

Summer Resident

WINTERING AREA: 5

HABITAT REQUIREMENTS: Grassy fens, moist grasslands, old fields with dense growth, dense cultivated grain fields, shallow marshes dominated by sedges. Nests interwoven with live grasses.

FEEDING: Feeds on insects and spiders by foraging on the ground and in low vegetation, gleaning prey and possibly hawking flying insects as well.

STATUS AND MANAGEMENT: Sedge Wrens do not appear frequently enough in the BBS data for the region to project long-term trends; however, numbers have declined significantly in North Dakota in the most recent survey years. Although the long-term trend for the wren's North American range in its entirety is relatively stable, there have been significant, dramatic declines in the most recent survey years, especially in the upper Midwest. Loss of freshwater wetlands likely has contributed to the increasing scarcity of this species. Sedge Wrens are nomadic, opportunistic breeders, with populations commonly shifting nesting areas between years. Known to destroy eggs of other small, marsh-nesting birds, including conspecifics.

FURTHER READING: Crawford, 1977; Picman and Picman, 1980.

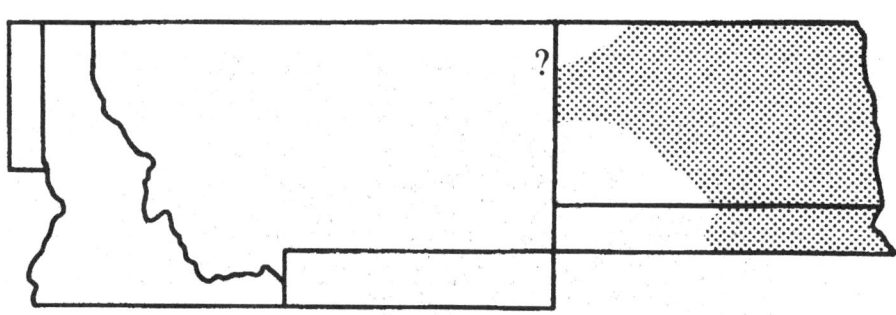

MARSH WREN

Cistothorus palustris
Troglodytidae

Troglodytidae

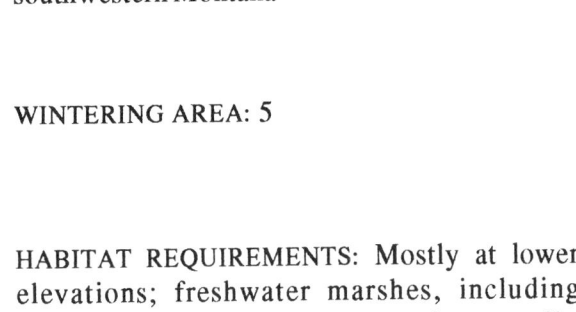

Summer Resident, occasionally Permanent Resident in western and southwestern Montana

WINTERING AREA: 5

HABITAT REQUIREMENTS: Mostly at lower elevations; freshwater marshes, including reservoirs, with emergent rushes, sedges, cattails, or tule to support over-water nests.

FEEDING: Feeds on insects, snails, and other small aquatic invertebrates. Forages by gleaning prey from the ground or foliage and by hawking flying insects.

STATUS AND MANAGEMENT: Numbers are small but apparently stable in Idaho and Montana. Marsh Wrens appear to be more abundant and increasing slightly in North Dakota with significant increase seen in the most recent survey year. Marsh Wrens are significantly increasing in the West as a whole, and appear to be stable or increasing slightly on a continent-wide basis. Favored nesting habitats in eastern portion of region are those with species of bulrush (Kantrud in Stewart, 1975). Known to destroy and sometimes consume eggs of other small, marsh-nesting birds, especially blackbirds and conspecifics.

FURTHER READING: Kroodsma, 1989; Leonard and Picman, 1986, 1987; Metz, 1991; Picman, 1984.

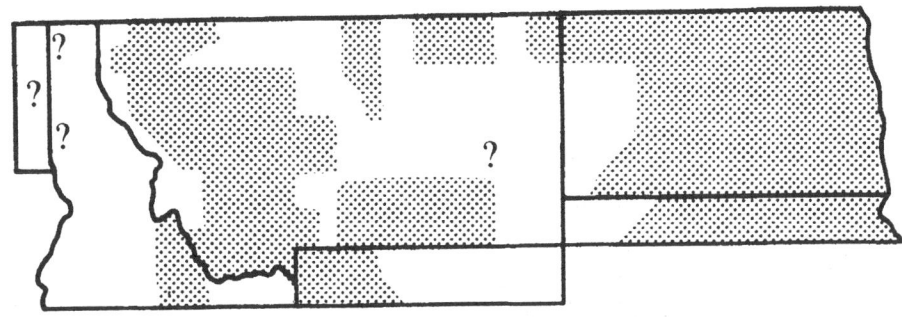

RUBY-CROWNED KINGLET

Regulus calendula
Sylviinae

Sylviinae

Summer Resident, rare Winter Resident in westernmost portion of region

WINTERING AREA: 4

HABITAT REQUIREMENTS: Primarily coniferous forests and woodlands, but also found in mixed coniferous-deciduous woodland. Apparently avoids pure or nearly-pure deciduous forest in central Rockies (Scott and Crouch, 1988b). Nests high in coniferous tree. In migration, also commonly uses riparian woodlands.

FEEDING: Takes insects and spiders in varied ways: forages in trees and shrubs gleaning items from leaves and bark, hovers and gleans, and hawks flying insects. Also feeds on tree sap at sapsucker wells, takes berries, and includes a few seeds in diet.

STATUS AND MANAGEMENT: Numbers show a slightly increasing trend in Idaho and appear to be stable in Montana, as elsewhere in the West. Continent-wide, the picture is mixed with overall stability as a result of declines in some areas being balanced by increases of the same magnitude elsewhere. Favors mature, dense coniferous forests for nesting. In contrast to the closely related Golden-crowned Kinglet, Ruby-crowns are generally more abundant in rotation-age forest than in old growth (Mannan and Meslow, 1984; Tobalske et al., 1991). Winter studies suggest that conditions experienced on the wintering grounds appear to control the size of subsequent breeding population. Breeding biology and breeding season ecology have been little studied.

FURTHER READING: Keast and Saunders, 1991.

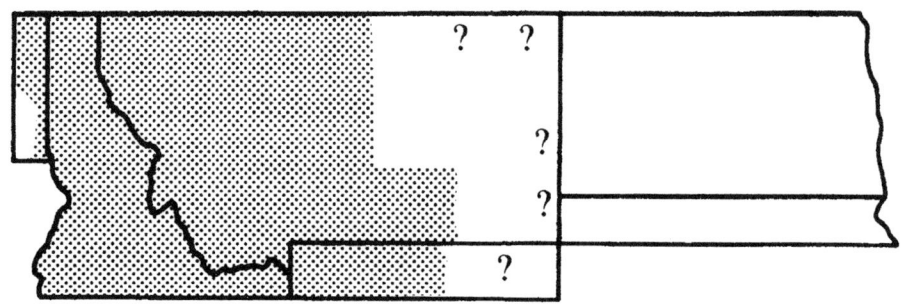

EASTERN BLUEBIRD

Sialia sialis
Muscicapinae

Summer Resident

WINTERING AREA: 6

HABITAT REQUIREMENTS: Forest edge, open woodland with grasslands, shelterbelts, riparian woodland, aspen woodland. Aside from nestboxes, nests most commonly in woodpecker-excavated cavity in standing snag, less frequently in live tree, fence post, or utility pole.

FEEDING: Diet composed primarily of insects, snails, earthworms, and other terrestrial invertebrates, and includes a substantial amount of small fruits, especially berries. Frequently hawks low-flying insects from low perch, in addition to gleaning from foliage.

STATUS AND MANAGEMENT: Numbers appear to be stable in North Dakota. Eastern Bluebirds appear too infrequently on western BBS routes to project a trend for western populations in general, but numbers for North America overall have stabilized with significant widespread increases in recent survey years. Listed as a Vulnerable Species in Canada. Sustained decline throughout the bluebird's range over the past several decades was attributed to competition with European Starlings and House Sparrows for nest cavities, competition with flocks of blackbirds and grackles for winter food supplies, impacts of pesticides, and severe winter weather; widespread successful campaign to promote use of nestboxes has led to recent increases in numbers.

FURTHER READING: Pinkowski, 1979; Plissner and Gowaty, 1988; Rendell and Robertson, 1990; Sauer and Droege, 1990; Zeleny, 1976.

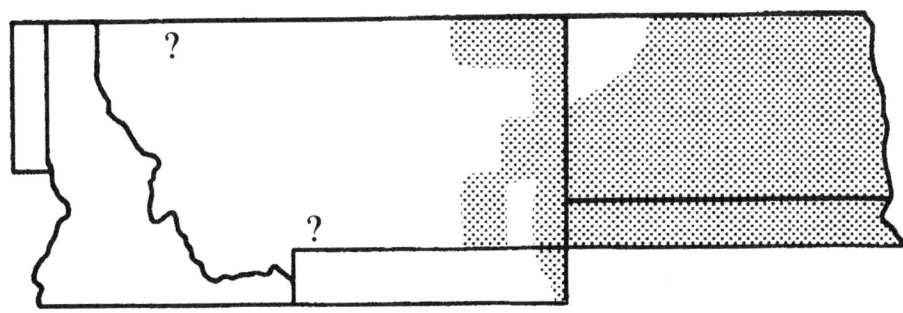

WESTERN BLUEBIRD

Sialia mexicana
Muscicapinae

Summer Resident

WINTERING AREA: 6

HABITAT REQUIREMENTS: Open coniferous or deciduous woodland with standing snags and interspersed patches of grasslands or adjacent montane meadows, riparian woodland, burned or logged areas with abundant standing snags. Nests in natural or woodpecker-excavated cavity, favoring aspens or ponderosa pines.

FEEDING: Diet composed primarily of insects, snails, earthworms, and other terrestrial invertebrates, and includes a substantial amount of small fruits, especially berries. Frequently hawks low-flying insects or sallies from low perch to capture prey on the ground, in addition to gleaning from foliage.

STATUS AND MANAGEMENT: Within the region, Western Bluebirds appear only on BBS routes in Idaho and numbers are very small but indicate a significantly declining trend. The overall picture throughout the range is one of small but continued decline. In comparison to the Eastern Bluebird, the western species have not received widely instituted nestbox programs aimed at stemming widespread, long-term declines in their numbers. Competition with European Starlings and House Sparrows for nest cavities has been invoked as a significant factor in the Western Bluebird's decline. Surprisingly little fieldwork has been conducted on breeding biology and ecology.

FURTHER READING: Mock, 1991; Mock et al., 1991.

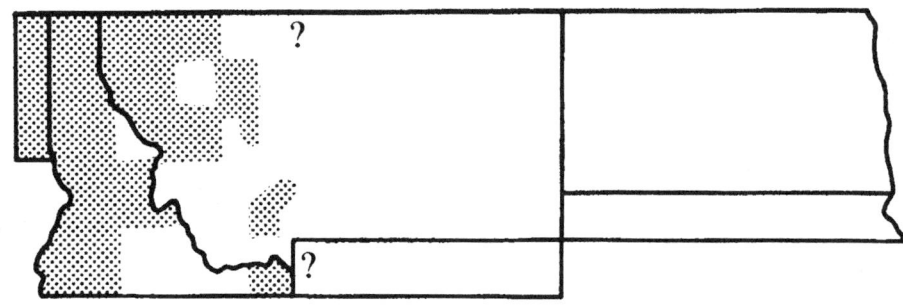

MOUNTAIN BLUEBIRD

Sialia currucoides
Muscicapinae

Summer Resident

WINTERING AREA: 5

HABITAT REQUIREMENTS: Open woodland of all types, forest edge, badlands, occurring across all elevations with suitable habitat. Nests in natural or woodpecker-excavated cavity in standing snags.

FEEDING: Diet composed primarily of insects, and includes few small fruits. Forages most frequently by swooping down from elevated perch to capture insects on the ground. Often hovers while gleaning items from the ground or from vegetation.

STATUS AND MANAGEMENT: Numbers have declined sharply in Idaho but appear to be stable in Montana, and possibly in North Dakota as well, although sample sizes are exceedingly small there. Mountain Bluebirds have sustained marked declines in recent years but appear now to have stabilized in most parts of their range. Within forested areas of the northern Rockies, nesting populations appear to depend heavily on burned areas with abundant standing snags (Hutto, in prep.). In comparison to the Eastern Bluebird, the western species have not received widely instituted nestbox programs, although they readily will nest in suitable nestboxes.

FURTHER READING: Herlugson, 1981; Power, 1980.

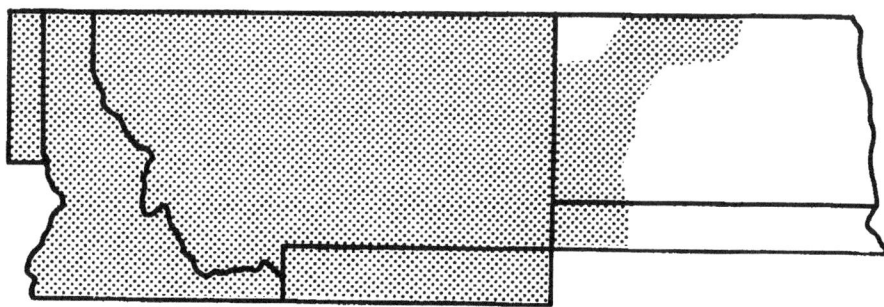

TOWNSEND'S SOLITAIRE

Myadestes townsendi
Muscicapinae

Permanent Resident but in many areas undergoes marked altitudinal migration to lower elevations in winter

WINTERING AREA: 5

HABITAT REQUIREMENTS: Montane and subalpine open coniferous forest, especially on slopes, steep rocky areas with shrub thickets; also open woodland and riparian woodland in winter. Nests on the ground in well-protected setting, frequently among tree roots or in brush pile.

FEEDING: Feeds on insects, spiders, worms, and other terrestrial invertebrates and includes substantial amount of small fruits, especially berries, in its diet. Often depends largely on berries, especially of juniper, in winter. Commonly forages by sallying from exposed perch in flycatcher-like fashion to capture insects in flight. Also gleans insects and fruits from vegetation and feeds on the ground.

STATUS AND MANAGEMENT: Populations are declining significantly in Idaho, and appear to be declining slightly in Montana. In the West as a whole, populations display a trend toward slightly increasing numbers. In winter, often defends territory encompassing a supply of berries or other small fruits. Breeding biology and ecology are little known.

FURTHER READING: Lederer, 1977; Salomonson and Balda, 1977.

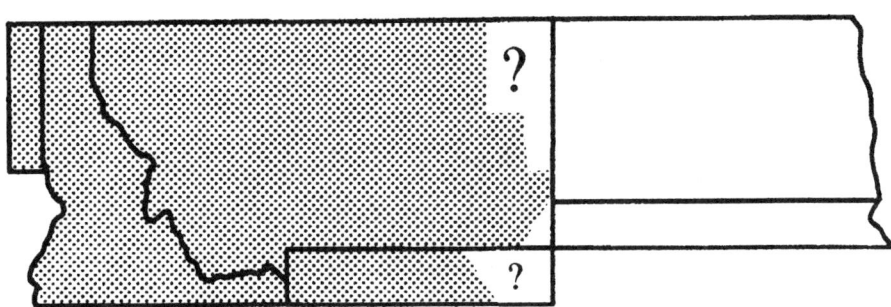

VEERY

Catharus fuscescens
Muscicapinae

Summer Resident

WINTERING AREA: 1

HABITAT REQUIREMENTS: Usually near water in moist, deciduous forest (especially aspen) with shrubby understory, also second growth, swamps, and riparian thickets. Nests on the ground, low in shrub, or on top of low stump.

FEEDING: Breeding season diet is almost exclusively composed of insects and spiders, only occasionally including a few other terrestrial invertebrates. Small fruits are consumed primarily in autumn. Usually feeds on the ground by picking items from the substrate or from vegetation; commonly feeds in bluebird-like fashion by swooping down from low perch to capture insects on the ground. Also gleans from foliage in low shrubs, but only rarely hawks flying insects.

STATUS AND MANAGEMENT: Numbers appear to be declining throughout the region, most significantly so in North Dakota. A slight declining trend is apparent throughout the West and continent-wide. A common cowbird host.

FURTHER READING: Bertin, 1977; Holmes and Robinson, 1988; Noon, 1981; Sousa, 1982.

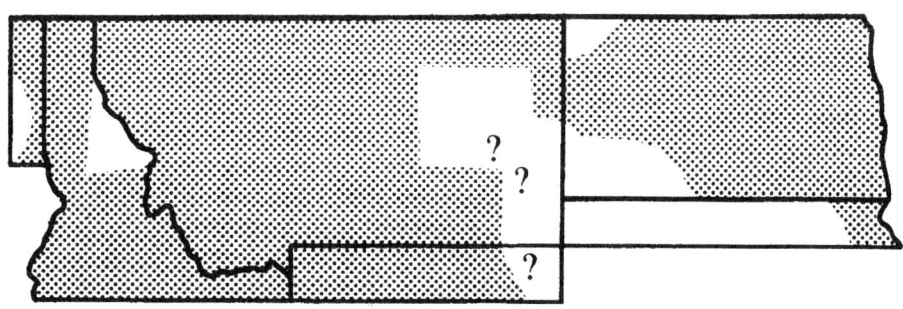

SWAINSON'S THRUSH

Catharus ustulatus
Muscicapinae

Muscicapinae

Summer Resident

WINTERING AREA: 3

HABITAT REQUIREMENTS: Mixed conifer old growth, especially with moist, wooded slopes, and shrubby understories, aspen forests with dense shrubs, swamps, and riparian thickets; also dense woodlands of all sorts in migration. Usually nests low in dense shrub, occasionally in small conifer tree, almost always near water.

FEEDING: Primarily consumes insects and spiders, only occasionally including other terrestrial invertebrates and small fruits. Forages on the ground to a lesser extent than the other ecologically similar ground thrushes; instead mostly forages by gleaning prey from foliage, by swooping down from low perch to capture prey on the ground, by gleaning while hovering, and by hawking flying insects.

STATUS AND MANAGEMENT: Regional picture is mixed with a declining trend in Idaho and a stable or slightly increasing trend in Montana. The overall pattern in the West is one of slight decline and one of relative stability continent-wide, although there have been significant decreases in the Canadian prairie provinces in the most recent survey years and extensive, widespread decline in the Sierra Nevada (Hejl et al., 1988). Swainson's Thrush is closely associated with old-growth mixed conifer forests (Finch and Reynolds, 1988; Hejl and Woods, 1991; Mannan and Meslow, 1984; Tobalske et al., 1991). Scattered declines have been linked, in part, to loss of Central American wintering habitat (Marshall, 1988); highly vulnerable to tropical deforestation (Morton, 1992).

FURTHER READING: Cherry, 1985; Holmes and Robinson, 1988; Noon, 1981; Sealy, 1974: Winker et al., 1992.

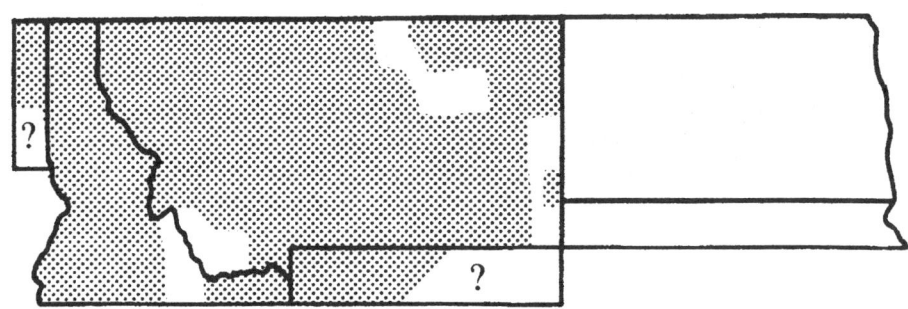

HERMIT THRUSH

Catharus guttatus
Muscicapinae

Summer Resident

WINTERING AREA: 4

HABITAT REQUIREMENTS: Moist coniferous or mixed coniferous-deciduous woodland is preferred but also found in drier spruce, ponderosa pine, or other coniferous woodland. Nests generally in low coniferous or deciduous tree, less frequently on the ground.

FEEDING: Takes insects, spiders, earthworms, small salamanders, and small fruits. Feeds primarily on the ground, but also gleans from foliage in shrubs and low trees, hovers to glean, and hawks flying insects.

STATUS AND MANAGEMENT: Slight declining trend in Idaho but a significantly increasing trend in Montana, which also is the pattern for the West as a whole with a generally increasing trend continent-wide. Significantly more abundant in old-growth Douglas fir/ponderosa pine than in rotation-age forest (Hejl and Woods, 1991; Mannan and Meslow, 1984). Nesting populations in old growth are sensitive to forest fragmentation (Keller and Anderson, 1992).

FURTHER READING: Holmes and Robinson, 1988; Martin and Roper, 1988; Noon, 1981; Sealy, 1974.

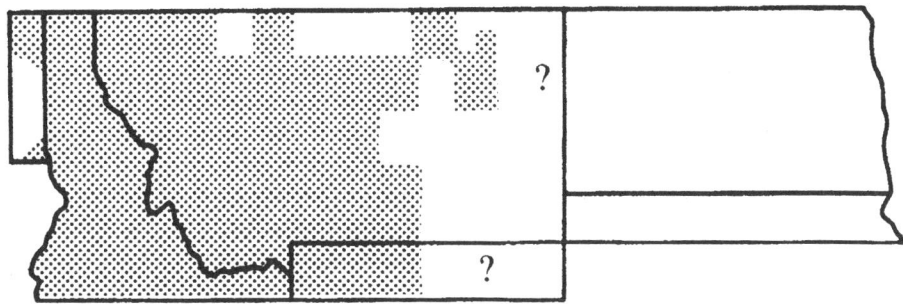

AMERICAN ROBIN

Turdus migratorius
Muscicapinae

Summer Resident, much less commonly a Permanent Resident, throughout the region

WINTERING AREA: 6

HABITAT REQUIREMENTS: Extremely widely distributed: open woodlands, forests, scrub, cultivated lands, grazed grasslands with scattered trees or shrubs, vicinity of human habitations. Usually nests in deciduous trees but also nests in conifers, shrubs, on human-built structures, and rarely on the ground. Need for mud in construction of nests results in their generally being built in the vicinity of water or moist ground.

FEEDING: Insects, earthworms, snails, and other invertebrates, as well as small fruits. Forages primarily on the ground, but also gleans from foliage.

STATUS AND MANAGEMENT: Numbers are increasing throughout the region, most significantly so in North Dakota; numbers increased significantly in Idaho in the most recent survey year. Overall, populations appear generally stable in the West, with a small but significant increasing trend seen continent-wide. American Robins expanded their range into the Great Plains and drier lowlands of the West as European-style agriculture extended into these areas and created suitable nesting and foraging sites. Responds positively to controlled burns in ponderosa pine woodland (Bock and Bock, 1983).

FURTHER READING: Hejl et al., 1988; Jung, 1992; Smith and Montgomerie, 1991; Weatherhead and McRae, 1990; Wheelwright, 1986.

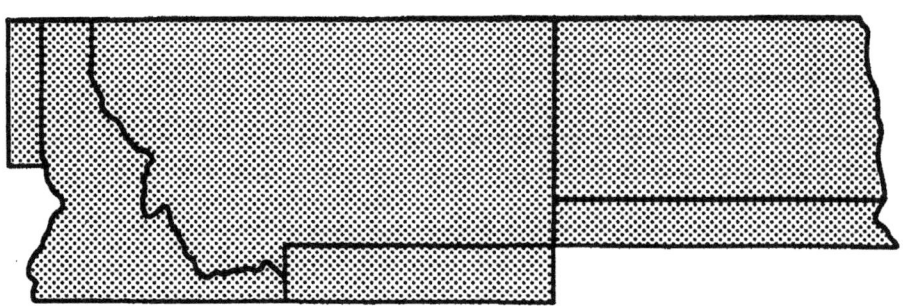

GRAY CATBIRD

Dumetella carolinensis
Mimidae

Summer Resident

WINTERING AREA: 4

HABITAT REQUIREMENTS: Low, dense thickets, usually in moist situations: often bordering woodland or in other open areas, riparian thickets, aspen woodland with shrub understory, mature shelterbelts, farmsteads and other areas of human habitations providing shrub thickets and scattered trees. Avoids coniferous forests and woodlands. Nests typically in dense shrubs.

FEEDING: Insects, spiders, and small fruits are taken by foraging on the ground and by gleaning from foliage.

STATUS AND MANAGEMENT: Numbers appear stable in Idaho but declining in Montana and North Dakota; however, populations exhibit an increasing trend in the West overall. Considered on a continent-wide basis, numbers appear stable.

FURTHER READING: Darley et al., 1977; Johnson and Best, 1982; Neudorf and Sealy, 1992; Yahner, 1991.

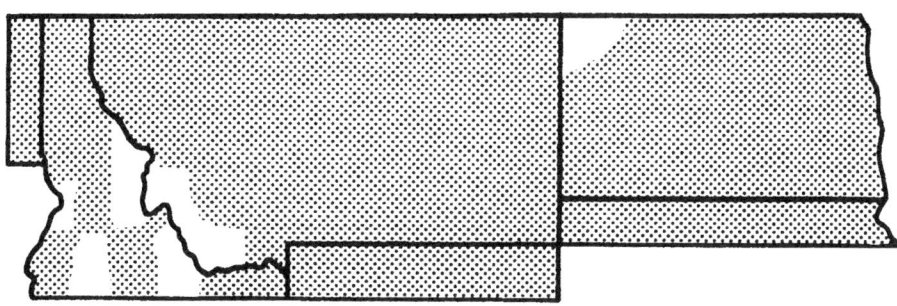

NORTHERN MOCKINGBIRD

Mimus polyglottos
Mimidae

Rare Summer Resident, even rarer Winter Resident

WINTERING AREA: 6

HABITAT REQUIREMENTS: Open and partly open habitats at lower elevations, mostly associated with farmsteads and other areas of human habitations providing shrub thickets and scattered trees interspersed with mowed lawns or meadows. Usually nests in shrubs, less frequently in low trees or vine tangles.

FEEDING: Takes a wide variety of insects and other terrestrial invertebrates as well as small fruits. Feeds primarily on the ground but also gleans from foliage.

STATUS AND MANAGEMENT: BBS data for this species are too few within the region to project population trends. In the West overall, numbers appear to be declining very slightly, and for the North American range as a whole there is a small but significant declining trend. Northern Mockingbirds have only recently expanded into the region.

FURTHER READING: Breitwisch et al., 1989; Derrickson, 1988; Logan, 1991; Zaias and Breitwisch, 1989.

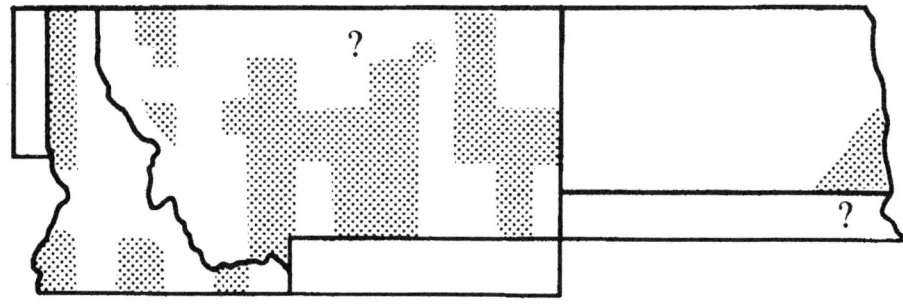

SAGE THRASHER

Oreoscoptes montanus
Mimidae

Summer Resident

WINTERING AREA: 5

HABITAT REQUIREMENTS: Mostly at lower elevations in shrubsteppe, almost invariably associated with sagebrush communities. Nests in shrub, usually sagebrush, or on ground beneath shrub. Selects nest sites in the tallest, most dense clumped shrubs, surrounded by little if any bare ground (Petersen and Best, 1991).

FEEDING: Diet composed of insects, other terrestrial invertebrates, and small fruits, especially berries. Forages on the ground and gleans foliage.

STATUS AND MANAGEMENT: Populations appear to be increasing in Idaho and Montana, as well as elsewhere generally throughout the West, although numbers declined significantly in Idaho in the most recent survey year.

FURTHER READING: Reynolds, 1981; Reynolds and Rich, 1978; Rich and Rothstein, 1985; Rotenberry and Wiens, 1989.

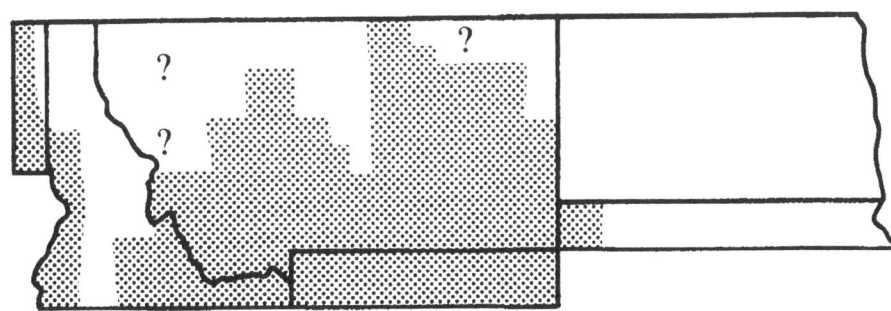

AMERICAN PIPIT

Anthus rubescens
Motacillidae

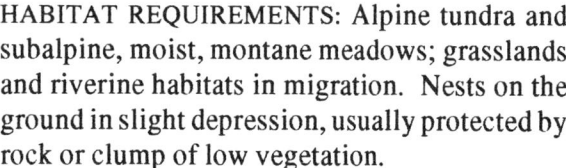

Summer Resident, rarely a Permanent Resident in westernmost portion of region

WINTERING AREA: 4

HABITAT REQUIREMENTS: Alpine tundra and subalpine, moist, montane meadows; grasslands and riverine habitats in migration. Nests on the ground in slight depression, usually protected by rock or clump of low vegetation.

FEEDING: Diet composed of insects, small aquatic invertebrates, seeds, and a few berries. Forages by gleaning items from the ground, occasionally hawks flying insects from the ground, and not averse to wading in shallow water to hunt small aquatic prey.

STATUS AND MANAGEMENT: No BBS data for this species are available from anywhere in the region. Very few BBS routes encompass alpine tundra habitat, hence the American Pipit has been recorded on only 10 BBS routes in the West. Although few in number, the indicated population trend is one of significant decline.

FURTHER READING: Hendricks, 1991; Hendricks and Norment, 1992; Miller and Green, 1987.

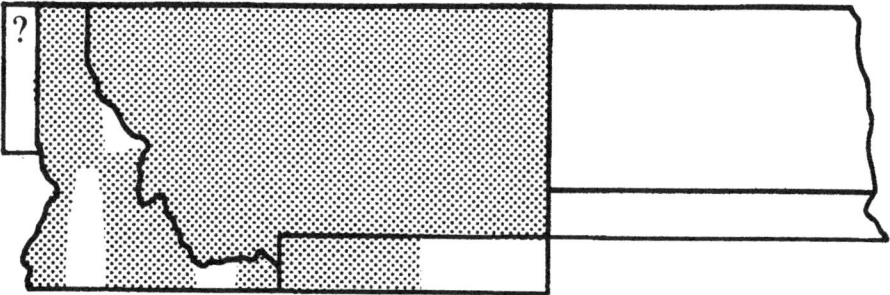

Transient throughout north central and eastern Montana

SPRAGUE'S PIPIT

Anthus spragueii
Motacillidae

Summer Resident

WINTERING AREA: 5

HABITAT REQUIREMENTS: Closely associated with extensive tracts of native shortgrass prairies, also alkaline meadows. Nests on the ground in clumps of grass or sedge.

FEEDING: Feeds on insects and seeds of grasses and forbs. Forages on the ground.

STATUS AND MANAGEMENT: Declining in both Montana and North Dakota, with significant declines over the species' range as a whole. Usually found only in ungrazed or very lightly grazed prairie; numbers have undergone a pronounced decline due to loss of native prairie habitat by conversion to agriculture and by habitat degradation resulting from livestock grazing. Breeding biology and ecology are little known.

FURTHER READING: Ehrlich, et al., 1988

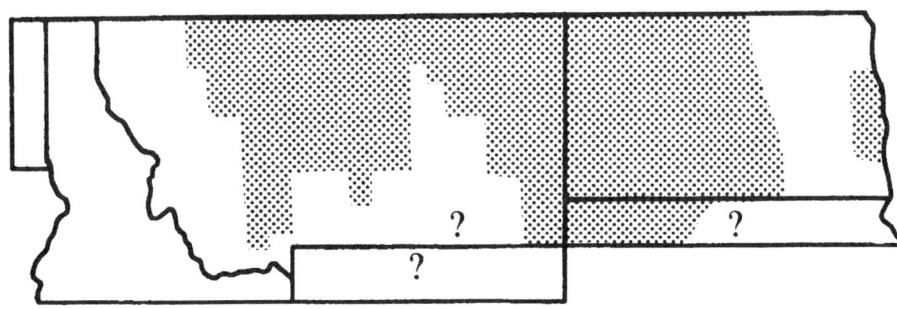

CEDAR WAXWING

Bombycilla cedrorum
Bombycillidae

Summer Resident from Idaho westward, Permanent Resident or Summer Resident from westernmost Montana eastward

WINTERING AREA: 4

HABITAT REQUIREMENTS: Open habitats providing berries: deciduous or coniferous woodland, forest edge, riparian woodland, farmsteads, shelterbelts, and any other open habitats (including around human habitations) with scattered trees. Usually nests in deciduous tree, less frequently in coniferous tree. Occasionally refurbishes and reuses old or abandoned waxwing nest.

FEEDING: One of the most fruit-dependent bird species in North America, taking mostly berries, but also consuming insects, flowers, and tree sap. Feeds primarily by gleaning from vegetation but also hawks flying insects.

STATUS AND MANAGEMENT: Numbers are declining slightly in Idaho but increasing in Montana and North Dakota. Populations are increasing significantly in the West as a whole, as well as continent-wide. Feeds gregariously in large winter flocks.

FURTHER READING: Leck and Cantor, 1979; McPherson, 1987; Mountjoy and Robertson, 1988a, b; Neudorf and Sealy, 1992; Rothstein, 1976a, b.

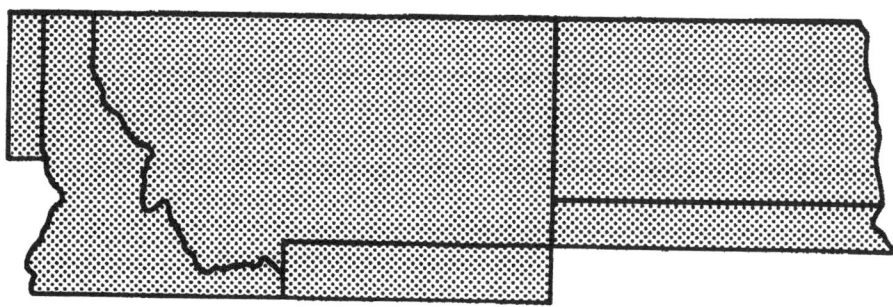

LOGGERHEAD SHRIKE

Lanius ludovicianus
Laniidae

Summer Resident, rarely a Winter Resident

WINTERING AREA: 5

HABITAT REQUIREMENTS: Open country from prairies to montane meadows providing scattered trees and shrubs, also pinyon/juniper woodland, shrubsteppe, arid scrub, croplands, shelterbelts. Nests in trees or shrubs, sometimes in tangle of vines. Occasionally refurbishes and reuses nests of other shrikes and of other passerines.

FEEDING: Primarily feeds on insects but also frequently takes small birds and occasionally mice or lizards, but only rarely consumes carrion. Swoops down onto prey from elevated perch, including fence posts or utility lines, or pursues birds in rapid, sustained flight, knocking them to the ground with a blow from the beak.

STATUS AND MANAGEMENT: Significantly declining in North Dakota but more or less stable in the rest of the region, although numbers have fluctuated greatly in Montana and Idaho. Western populations as a whole have undergone sustained significant declines, and the picture is no better in the East as the species has declined significantly on a continent-wide basis. The eastern subspecies is a candidate for federal Threatened or Endangered status in the U.S. In eastern Canada, the Loggerhead Shrike is listed as Endangered, while elsewhere throughout its range in Canada it is listed as a Threatened Species. Habitat loss and pesticide contamination are thought responsible for widespread declines, but there is little real understanding of the factors responsible. Recent studies in the upper Midwest and in the Southeast implicate problems associated with survival on the wintering grounds as being responsible for continued declines in breeding populations (Brooks and Temple, 1990; Gawlik and Bildstein, 1990), although site fidelity to breeding territories may be unusually low in Loggerhead Shrikes, thus leading to inherently low return rates (Haas and Sloane, 1989).

FURTHER READING: Fraser and Luukkonen, 1986; Morrison, 1980; Porter et al., 1975.

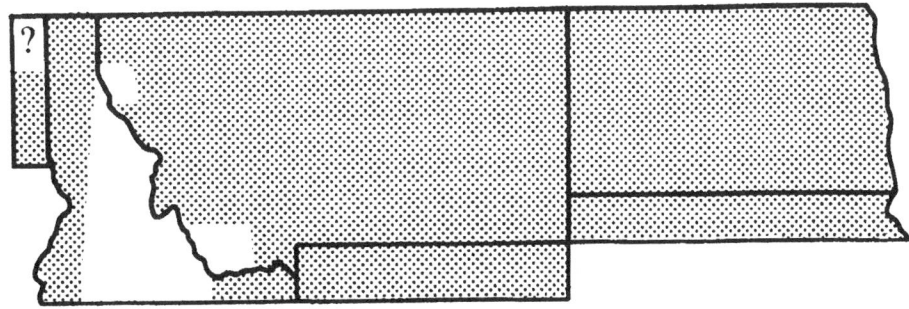

BELL'S VIREO

Vireo bellii
Vireonidae

Summer Resident

WINTERING AREA: 3

HABITAT REQUIREMENTS: Dense, riparian thickets of shrubs or small trees. Nests in dense shrub or low tree.

FEEDING: Nearly exclusively insects with a few berries added to the diet in late summer. Feeds by gleaning from foliage.

STATUS AND MANAGEMENT: Bell's Vireo reaches the northernmost point of its breeding range in North Dakota where it is too rare to have been encountered on regional BBS routes. Throughout its range, the species is recorded too infrequently on BBS routes to provide statistically meaningful population projections, other than to indicate that it continues in widespread decline.

The subspecies of Bell's Vireo found in California is federally listed as an Endangered Species, having disappeared from all but 5 percent of its former U.S. range. Widespread decline throughout the vireo's range has been linked closely with destruction of riparian habitats through channelization for flood control and agriculture and degradation of riparian habitats as a result of livestock grazing. A further important factor has been heavy nest parasitism by cowbirds.

FURTHER READING: Baird and Rieger, 1989; Barlow, 1962; Franzreb, 1989; Greaves, 1989; Hendricks and Rieger, 1989; Kus and Miner, 1989; Olson and Gray, 1989.

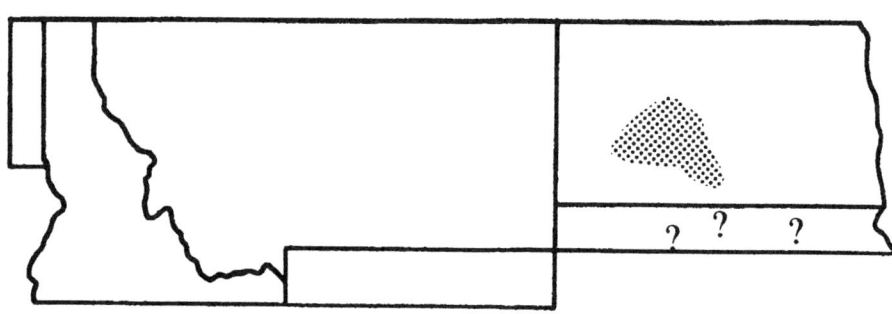

SOLITARY VIREO

Vireo solitarius
Vireonidae

Summer Resident

WINTERING AREA: 4

HABITAT REQUIREMENTS: Open woodlands with shrub understories: mixed coniferous-deciduous or coniferous woodland, montane woodland, and pine-dominated woodland. Nests are usually placed low in deciduous or coniferous trees.

FEEDING: Nearly exclusively insects with a few berries added to the diet in late summer or autumn. Feeds by gleaning from foliage, hawking flying insects, and gleaning from the bark surfaces of branches.

STATUS AND MANAGEMENT: Apparently declining in Idaho and Montana, but stable in the West as a whole and increasing significantly when data from all BBS routes are combined continent-wide, although behavioral data indicate that the eastern and western North American populations may in fact be separate species. Favors rotation-age rather than old-growth forest (Hejl and Woods, 1991) but is apparently sensitive to forest fragmentation (Aney, 1984). Responds positively to controlled burning of ponderosa pine forest or woodland (Bock and Bock, 1983). A common cowbird host and may suffer very heavy rates of parasitism in some western localities (Marvil and Cruz, 1989).

FURTHER READING: Barclay, 1977; James, 1978.

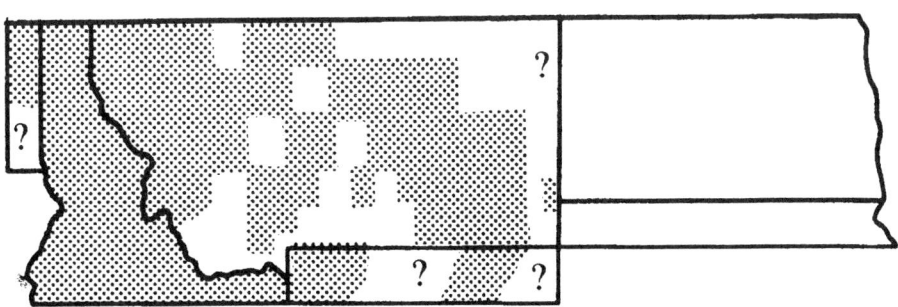

YELLOW-THROATED VIREO

Vireo flavifrons
Vireonidae

Summer Resident

WINTERING AREA: 4

HABITAT REQUIREMENTS: Mature deciduous forests and riparian woodland. Nests in deciduous trees.

FEEDING: Nearly exclusively insects with a few berries added to the diet in autumn. Feeds by gleaning from foliage.

STATUS AND MANAGEMENT: This species does not occur in the BBS database for North Dakota. In its North American range as a whole, the vireo has declined steadily, although numbers appear to have stabilized in the most recent survey years. The Yellow-throated Vireo is an eastern species that reaches the northwestern edge of its breeding range in North Dakota. Population declines in New England suburban habitats have been linked to pesticide spraying of deciduous trees. Extremely vulnerable to tropical deforestation (Morton, 1992).

FURTHER READING: James, 1978; Smith et al., 1978.

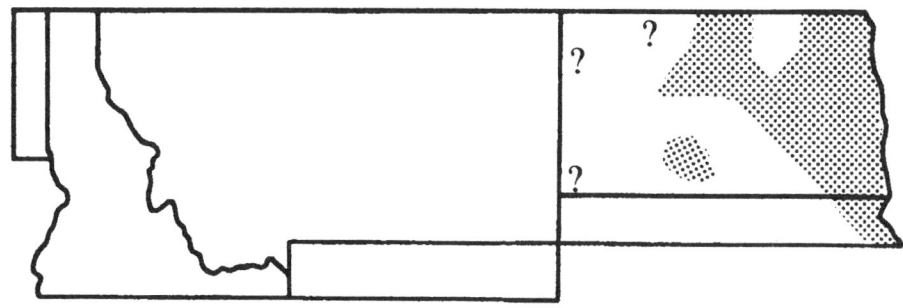

RED-EYED VIREO

Vireo olivaceus
Vireonidae

Summer Resident

WINTERING AREA: 1

HABITAT REQUIREMENTS: Aspen, poplar, or other deciduous forest and woodland, riparian woodland, well-treed areas in the vicinity of human habitations. Nests in dense shrubs or deciduous trees.

FEEDING: Nearly entirely insects with a few other terrestrial invertebrates taken occasionally and berries added opportunistically. Feeds most frequently by gleaning from vegetation while hovering, but also forages by gleaning from foliage.

STATUS AND MANAGEMENT: Infrequently encountered on Idaho BBS routes but apparently declining there, and declining in Montana and North Dakota, as well. Following steady declines throughout their range, Red-eyed Vireos now appear to be stable in the West as a whole and increasing slightly when viewed continent-wide, although significant declines in several parts of eastern North America were reported in the most recent survey years. Spraying of hardwood forest with pesticides to control gypsy moths necessitated a tripling in size of foraging areas by individual Red-eyed Vireos and caused a substantial shift in diet as a result of reduced caterpillar densities, but reproductive impacts have not been assessed (Cooper et al., 1990). One of the species most frequently parasitized by cowbirds. Highly vulnerable to tropical deforestation (Morton, 1992).

FURTHER READING: Barlow and Rice, 1977; Darveau et al., 1992; Graham, 1988; Pletschet, 1987; Robinson, 1981; Williamson, 1971.

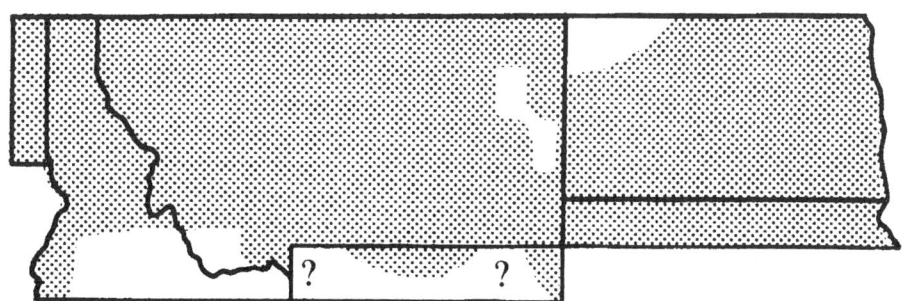

WARBLING VIREO

Vireo gilvus
Vireonidae

Summer Resident

WINTERING AREA: 3

HABITAT REQUIREMENTS: Typically found in mature riparian woodland, but also uses open deciduous forests and woodlands, aspen or birch groves within coniferous forest, mature shelterbelts and well-wooded farmsteads. Nests in deciduous trees or shrubs.

FEEDING: Almost entirely insects and spiders plus a few berries. Feeds by gleaning from foliage and by gleaning from vegetation while hovering.

STATUS AND MANAGEMENT: Populations appear to be increasing throughout the region, significantly so in North Dakota. Increasing significantly in the West as a whole, which drives an overall trend of significant increase continent-wide, although there have been significant declines in several parts of the eastern range in the most recent survey years. Reduction in extent of uncut aspen stands and conifer invasion of aspen forest are likely to negatively impact vireo populations (Finch and Reynolds, 1988). Population declines in suburban areas within the eastern North American range have been linked to pesticide spraying of deciduous trees. Common cowbird host.

FURTHER READING: Howes-Jones, 1985; James, 1976; Pletschet, 1987.

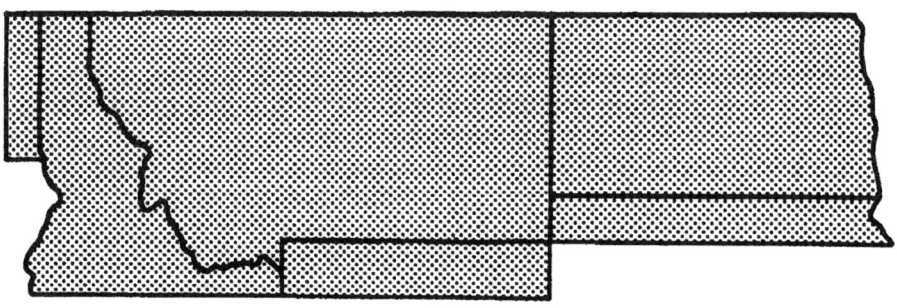

TENNESSEE WARBLER

Vermivora peregrina
Parulinae

Summer Resident

WINTERING AREA: 3

HABITAT REQUIREMENTS: Deciduous or mixed deciduous-coniferous woodland with aspen or poplar, boggy thickets. Nests on the ground, usually at base of low shrub or in sphagnum hummock (in boggy areas).

FEEDING: Largely insects but includes some small berries. Forages most often at the ends of branches by gleaning from foliage and bark.

STATUS AND MANAGEMENT: Appears too infrequently in the BBS database for the region to project population trends. The trend in the West as a whole is one of slight increase, as is the case also on a continent-wide basis, although there has been a widespread declining trend in the most recent survey years. Tennessee Warblers are spruce budworm specialists—populations respond rapidly to budworm outbreaks and crashes, resulting in marked regional fluctuations in numbers from year to year.

FURTHER READING: Morse, 1989; Quay, 1989; Winker et al., 1991.

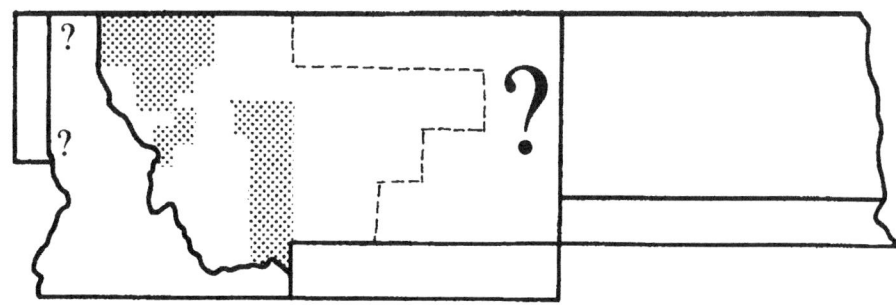

Transient eastward to dotted line

ORANGE-CROWNED WARBLER

Vermivora celata
Parulinae

Summer Resident

WINTERING AREA: 4

HABITAT REQUIREMENTS: Riparian thickets and woodland, open brushy woodland, shrubby, post-fire communities, aspen groves, forest edge. Nests usually or exclusively on the ground.

FEEDING: Takes insects, fruit, and floral nectar by gleaning from foliage and branches. Also consumes tree sap at sapsucker wells.

STATUS AND MANAGEMENT: Numbers appear to be declining in Idaho and Montana, mirroring an overall declining trend in the West and continent-wide, although overall numbers increased significantly in the most recent survey year. In some areas, this warbler is associated exclusively with aspen and mixed aspen/conifer stands (Finch and Reynolds, 1988). Although widespread, many details of breeding biology are unknown.

FURTHER READING: Foster, 1969; Morrison, 1981; Morse, 1989.

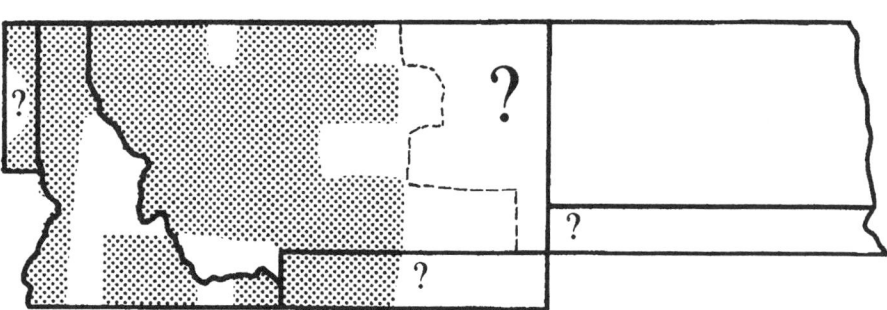

NASHVILLE WARBLER

Vermivora ruficapilla
Parulinae

Summer Resident

WINTERING AREA: 4

HABITAT REQUIREMENTS: Open deciduous woodland or deciduous groves in coniferous woodland, riparian woodland, second growth, forest-edged bogs; in all cases, prefers areas with shrub undergrowth. Nests on the ground, generally beneath a shrub.

FEEDING: Diet is composed wholly of insects, which are captured by gleaning from foliage and branches and by hovering while gleaning and occasionally foraging on the ground.

STATUS AND MANAGEMENT: Within the region, frequency of occurrence on BBS routes has been too little and pattern of occurrence has been too erratic to project meaningful population trends. Overall pattern in the West has been a trend of small but significant increase, with a slight increasing trend seen continent-wide, as well. Across its range overall, the Nashville Warbler is most closely associated with second-growth deciduous woodlands. Riparian populations are impacted negatively by livestock (Mosconi and Hutto, 1982).

FURTHER READING: Johnson, 1976; Morse, 1989.

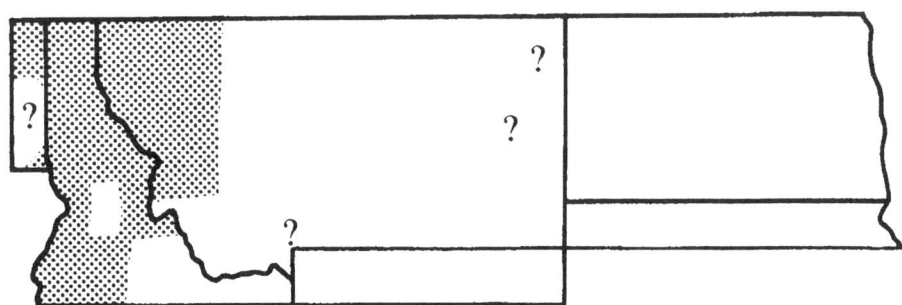

YELLOW WARBLER

Dendroica petechia
Parulinae

Summer Resident

WINTERING AREA: 4

HABITAT REQUIREMENTS: Primarily associated with riparian thickets (especially of willows) and riparian woodland with dense understories; also found in shelterbelts, well-wooded areas of towns and farmsteads. Nests in dense shrub or in small deciduous tree.

FEEDING: Diet is composed mostly of insects but also will include a few berries in diet. Forages primarily by foliage gleaning and also gleans from bark of branches, hawks flying insects, and hovers to glean from vegetation.

STATUS AND MANAGEMENT: Numbers appear to be declining in Idaho, stable in Montana, and declining in North Dakota. In the west as a whole, populations appear to have stabilized following sharp declines in recent decades. The continent-wide trend indicates small but significant increase, although numbers have declined significantly in the far West in the most recent survey years. The Yellow Warbler is one of the three species most frequently parasitized by cowbirds and has the largest breeding range of any wood warbler. Western populations have declined markedly in response to loss of riparian thickets due to a combination of drought, channelization for flood control and agriculture, and impacts from livestock grazing. Populations respond well to willow restoration and regeneration that occur when riparian areas are protected from cattle (Taylor and Littlefield, 1986).

FURTHER READING: Graham, 1988; Schroeder, 1982a; Sealy, 1992; Sealy et al., 1989; Weatherhead, 1989; Wiedenfeld, 1992.

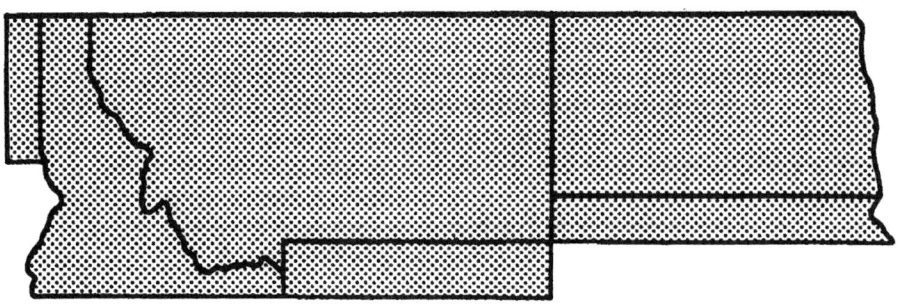

YELLOW-RUMPED WARBLER

Dendroica coronata
Parulinae

Summer Resident, rare Permanent Resident in western Montana and Idaho

WINTERING AREA: 4

HABITAT REQUIREMENTS: Open to dense, montane coniferous forest. There is also a small breeding population in southwestern North Dakota in an isolated tract of ponderosa pine forest. Nests in coniferous trees.

FEEDING: Primarily insects but diet regularly includes waxy berries, especially outside of the breeding season. Forages primarily by gleaning from foliage but also commonly hawks flying insects and hovers while gleaning from vegetation.

STATUS AND MANAGEMENT: Populations appear to be increasing significantly in Idaho, but declining slightly in Montana, with no data from North Dakota. In the West overall, numbers appear to be stable, while continent-wide there is a small but significantly increasing trend. Significantly more abundant in old-growth than in rotation-age forests (Hejl and Woods, 1991), and significantly more abundant in unfragmented compared to fragmented forests (Keller and Anderson, 1992). Responds positively to prescribed burning of open ponderosa pine forest and woodland (Bock and Bock, 1983). This is arguably the most abundant North American warbler.

FURTHER READING: Morse, 1989; Place and Stiles, 1992.

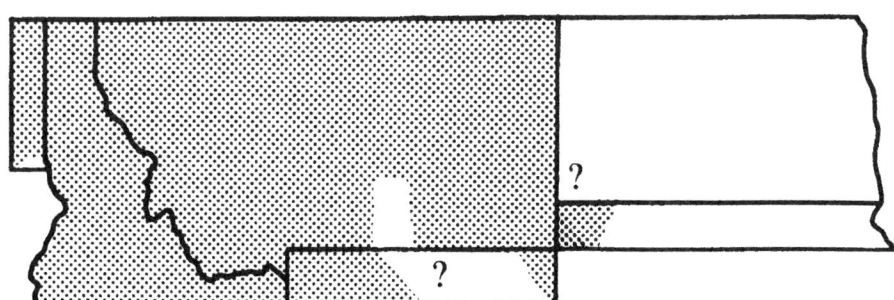

TOWNSEND'S WARBLER

Dendroica townsendi
Parulinae

Summer Resident

WINTERING AREA: 4

HABITAT REQUIREMENTS: Mature coniferous and mixed coniferous-deciduous forest. Nests in conifer trees. In old-growth Douglas fir/ponderosa pine in northeastern Oregon, nest sites are associated with high canopy volumes of grand and Douglas firs (Mannan and Meslow, 1984).

FEEDING: Largely or entirely insects, but may include plant galls and a few seeds. Generally forages high in trees by gleaning items from the vegetation and occasionally hawking flying insects.

STATUS AND MANAGEMENT: Within the region, frequency and pattern of occurrence on BBS routes are erratic; numbers appear to be declining in Idaho and increasing in Montana, but it is chancy at best to attempt projections from these data. Overall pattern in the West is a trend of slight increase in numbers. Consistently more abundant in old-growth coniferous forests and sensitive to forest fragmentation (Hejl and Woods, 1984; Mannan and Meslow, 1984; Tobalske et al., 1991). Townsend's Warbler should be considered a forest-interior nest-ing species. Breeding biology has been little studied.

FURTHER READING: Jackson et al., 1992; Morrison, 1983; Morse, 1989.

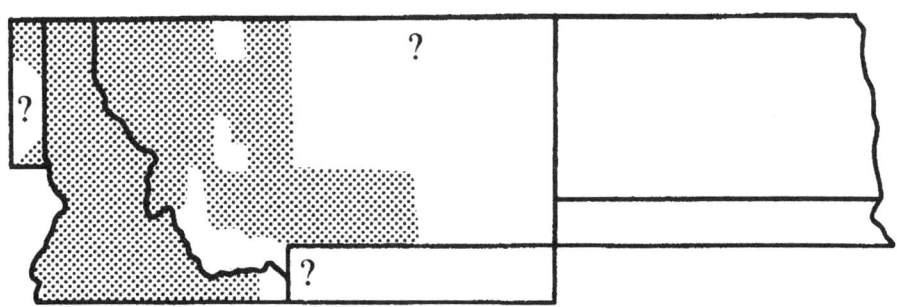

BLACK-AND-WHITE WARBLER

Mniotilta varia
Parulinae

Summer Resident

WINTERING AREA: 4

HABITAT REQUIREMENTS: Open deciduous woodland, mixed woodland in badlands, second-growth deciduous woodland. Nests on the ground, usually well concealed by low branches of shrub.

FEEDING: Consumes insects gleaned from bark of tree branches and trunks in a nuthatch-like foraging manner, but also capable of gleaning from foliage.

STATUS AND MANAGEMENT: Numbers appear to be declining in North Dakota but sample sizes are very small. Sample sizes for the West as a whole are likewise very small but numbers appear to be relatively stable. This is primarily an eastern species, and numbers on a continent-wide basis appear stable with significant increases in several areas in the most recent survey years. The Black-and-white Warbler is considered to be a forest-interior nesting species. Highly vulnerable to tropical deforestation. (Morton, 1992). Frequent cowbird host.

FURTHER READING: Morse, 1989.

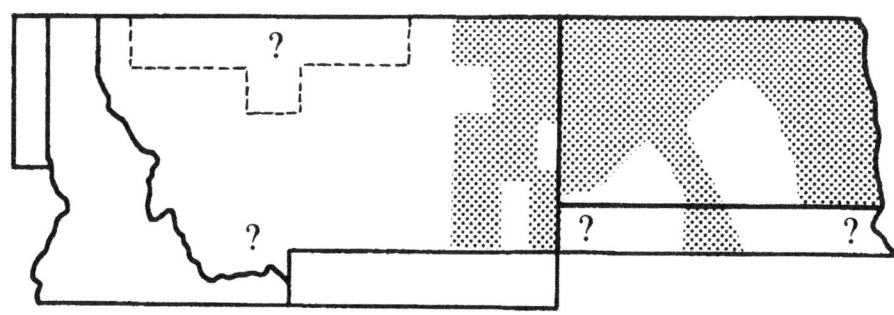

AMERICAN REDSTART

Setophaga ruticilla
Parulinae

Summer Resident

WINTERING AREA: 3

HABITAT REQUIREMENTS: Mature deciduous forest, open deciduous (including riparian) and mixed deciduous-coniferous woodland, forest edge, second growth, and aspen groves. Nests in deciduous tree or shrub, occasionally using abandoned nests of other small songbirds.

FEEDING: Forages on insects primarily by hovering while gleaning from foliage, also hawks flying insects and gleans from vegetation while moving through trees and shrubs.

STATUS AND MANAGEMENT: Populations exhibit a declining trend in Montana, the only part of the region with reasonable sample sizes from the BBS database; numbers appear to be stable in Idaho and increasing slightly in North Dakota, but sample sizes are exceedingly small for both states. The trend in the West overall as well as continent-wide is one of relative stability, although there was a sharp, significant decline in continent-wide numbers in the most recent survey years. There is strong evidence that the size of some breeding populations is determined by impacts of nest predators and brood parasites on the breeding grounds (Sherry and Holmes, 1992), while other populations may be most limited by conditions encountered on the wintering grounds (Bennett, 1980).

FURTHER READING: Lemon et al., 1992; Maurer and Whitmore, 1981; Morse, 1989; Procter-Gray, 1991; Sabo and Holmes, 1983; Secunda and Sherry, 1991; Sherry and Holmes, 1988.

OVENBIRD

Seiurus aurocapillus
Parulinae

Summer Resident

WINTERING AREA: 4

HABITAT REQUIREMENTS: Mature forests: floodplain deciduous forests, upland deciduous or (less frequently) mixed deciduous-coniferous forests on north-facing slopes. Nests on the ground.

FEEDING: Diet is composed mostly of insects supplemented with spiders, snails, earthworms, and other terrestrial invertebrates. Forages on the ground, picking items from the substrate or in leaf litter, only occasionally gleaning from foliage or bark in shrubs and saplings.

STATUS AND MANAGEMENT: Sample sizes are small within the region but numbers appear to be stable or increasing slightly in Montana and North Dakota. Similarly, the number of BBS routes with Ovenbirds in the West overall is small but numbers appear to be stable. Based on a much larger sample size, the continent-wide pattern also appears to be one of stability, although some areas in eastern North America exhibited significant declines in the most recent survey years while other areas experienced significant increases. Ovenbird populations are sensitive to forest fragmentation and the species is considered to be a forest-interior nesting species. Highly vulnerable to tropical deforestation (Morton, 1992). Frequent cowbird host.

FURTHER READING: Holmes and Robinson, 1988; Morse, 1989; Smith and Shugart, 1987; Zach and Falls, 1975.

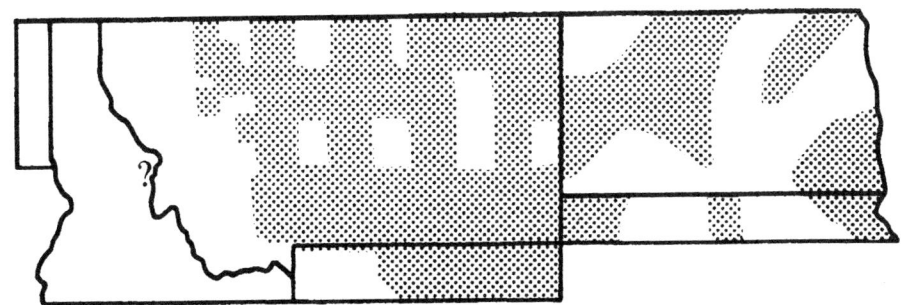

NORTHERN WATERTHRUSH

Seiurus motacilla
Parulinae

Summer Resident

WINTERING AREA: 4

HABITAT REQUIREMENTS: Montane woodlands with bogs or other standing water, wooded riparian swamps and second-growth swamp-forests, wooded edges of ponds, lakes, and streams. Nests close to water, on the ground amid roots of upturned trees, or beneath overhang in streambank.

FEEDING: Feeds on small aquatic invertebrates, including insects, and on terrestrial insects and other invertebrates. Forages on the ground, usually at water's edge or wading in shallow water, often flipping through leaf litter and pulling dead leaves from the water to inspect them for adherent prey. Also gleans from low foliage and occasionally hawks flying insects from the ground.

STATUS AND MANAGEMENT: Numbers appear to be stable or increasing in Idaho and Montana, but sample sizes are very small from Idaho and rather erratic from Montana. Western populations taken as a whole show a significantly increasing trend, while the pattern continent-wide is one of apparent slight increase.

FURTHER READING: Craig, 1987; Winker et al., 1992.

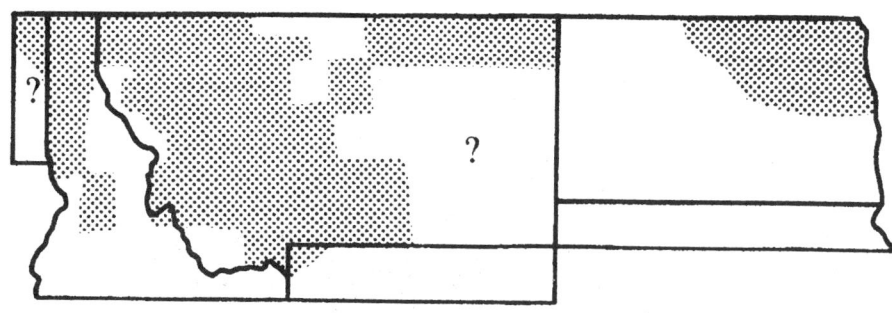

MacGILLIVRAY'S WARBLER

Oporornis tolmiei
Parulinae

Summer Resident

WINTERING AREA: 3

HABITAT REQUIREMENTS: Riparian thickets (especially willow), dense undergrowth of coniferous forest and forest edge, moist, brushy, montane slopes. Nests low in dense shrub or on the ground.

FEEDING: Entirely insects so far as known, although sap from sapsucker wells is taken when available in willows. Forages primarily by gleaning foliage low in shrubs but also gleans from bark of branches and feeds on the ground, as well.

STATUS AND MANAGEMENT: Numbers appear to be declining in Idaho but relatively stable or slightly increasing in Montana. Overall pattern in the West is one of slight decline. A riparian specialist, exhibiting very narrow habitat tolerance compared with most other insectivorous birds of riparian habitats (Douglas et al., 1992; Dobkin et al., in prep.). Riparian populations are adversely affected by livestock grazing (Mosconi and Hutto, 1982). Breeding biology has been little studied. An uncommon cowbird host.

FURTHER READING: Hutto, 1981a, b; Morrison, 1981.

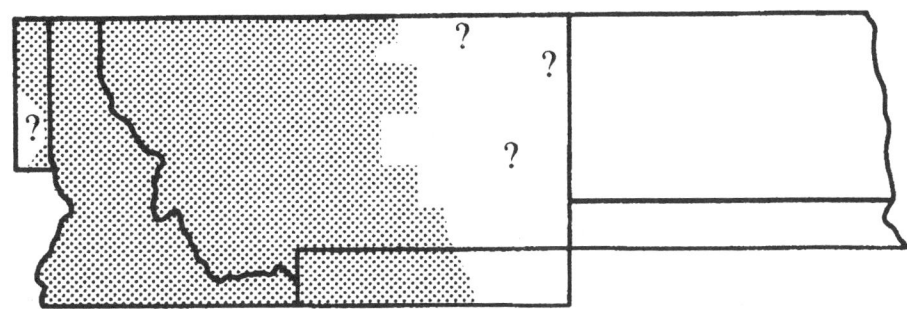

COMMON YELLOWTHROAT

Geothlypis trichas
Parulinae

Summer Resident

WINTERING AREA: 4

HABITAT REQUIREMENTS: Wet or at least moist habitats: montane willow thickets around beaver ponds, marshy areas, riparian thickets (especially willow), wet meadows, wooded wetland borders, shelterbelts, partially wooded residential areas. Nests low in shrub.

FEEDING: Diet composed of insects and spiders, with few other small invertebrates. Forages mostly in brush and low shrubs by gleaning from foliage and bark, also hovers to glean from vegetation, and hawks flying insects.

STATUS AND MANAGEMENT: Numbers appear to be increasing in Idaho but declining in Montana and North Dakota, with significant declines in the most recent survey year in North Dakota. In the West as a whole, populations show a significant increasing trend in contrast to a continent-wide declining trend, especially so in the most recent survey year, with significant declines in several parts of the eastern U.S. and in British Columbia. In North Dakota, favors wet meadows dominated by prairie cordgrass adjacent to streams and ponds. Within riparian habitats of the region, exhibits very narrow range of habitat selection compared with most other insectivorous birds (Douglas et al., 1992). Riparian populations are impacted adversely by livestock (Mosconi and Hutto, 1982; Sedgwick and Knopf, 1987). One of the three most frequent cowbird hosts.

FURTHER READING: Atwood, 1992; Hutto, 1981b; Lewis, 1972; Morimoto and Wasserman, 1991; Ritchison, 1991.

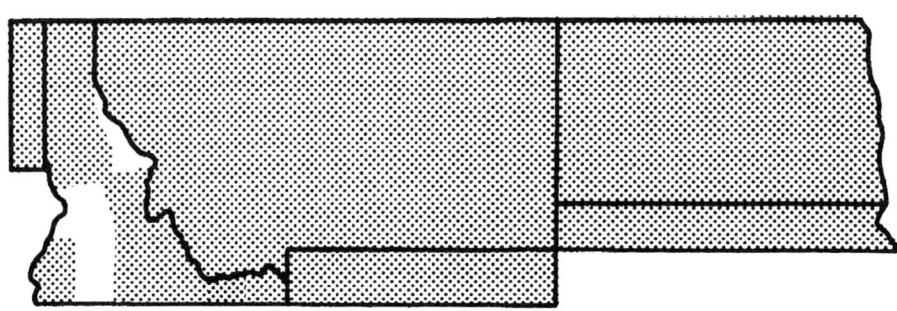

WILSON'S WARBLER

Wilsonia pusilla
Parulinae

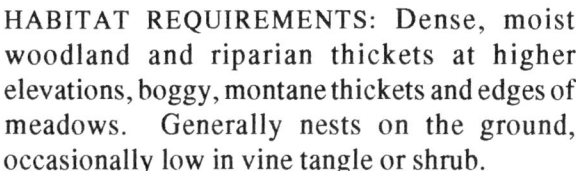

Summer Resident

WINTERING AREA: 4

HABITAT REQUIREMENTS: Dense, moist woodland and riparian thickets at higher elevations, boggy, montane thickets and edges of meadows. Generally nests on the ground, occasionally low in vine tangle or shrub.

FEEDING: Consumes primarily insects but also takes a few berries. Forages by gleaning from foliage and twigs, sometimes hovering to glean, and by hawking flying insects.

STATUS AND MANAGEMENT: Idaho populations exhibit a sharply and significantly declining trend, while Montana numbers also appear to be declining. In the West as a whole, however, numbers appear to be relatively stable, as is the case for the continent-wide trend, although numbers have declined in the most recent survey years. Relatively specialized in the types of riparian habitats utilized (Douglas et al., 1992). Riparian populations are negatively affected by livestock (Knopf et al., 1988). An uncommon cowbird host.

FURTHER READING: Morrison, 1981; Raley and Anderson, 1990; Stewart et al., 1977.

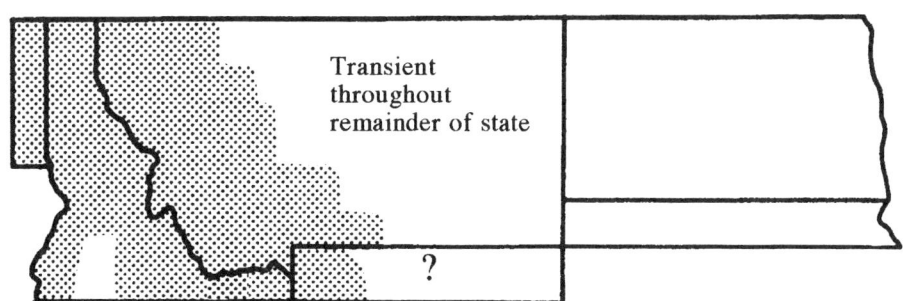

145

YELLOW-BREASTED CHAT

Icteria virens
Parulinae

Summer Resident

WINTERING AREA: 4

HABITAT REQUIREMENTS: Riparian thickets, second growth and scrub along swamp margins, dense shrub thickets and mountain mahogany woodlands on moist slopes. Nests in dense shrub.

FEEDING: Takes nearly equal proportions of insects and berries. Forages by gleaning from foliage within vegetation.

STATUS AND MANAGEMENT: Numbers have increased significantly in Idaho, declined slightly in Montana, and increased slightly in North Dakota. Western populations overall exhibit slight increase, while the pattern continent-wide is one of slight decrease with strong, significant decreases in the most recent survey years. Riparian populations are affected negatively by livestock (Sedgwick and Knopf, 1987). Frequently parasitized by cowbirds.

FURTHER READING: Ritchison, 1988; Thompson, 1977; Thompson and Nolan, 1973.

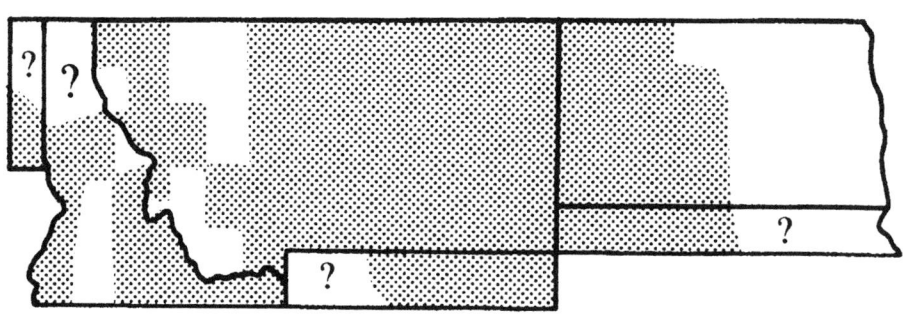

SCARLET TANAGER

Piranga olivacea
Thraupinae

Summer Resident

WINTERING AREA: 1

HABITAT REQUIREMENTS: Mature deciduous forest of floodplains and upland slopes. Nests in deciduous trees.

FEEDING: Insects, spiders, berries, and other fruits. Forages most frequently by gleaning from foliage while hovering, also gleans from foliage and bark while moving through trees, and hawks flying insects (more frequently by females than by males).

STATUS AND MANAGEMENT: North Dakota is the westernmost edge of the breeding range, but Scarlet Tanagers do not appear in the region's BBS database. Continent-wide, numbers have been fairly stable although steep, significant declines occurred in the most recent survey years. A relatively small wintering range translates into high vulnerability to tropical deforestation (Morton, 1992). Common cowbird host.

FURTHER READING: Holmes, 1986; Shy, 1984; Villard et al., 1992.

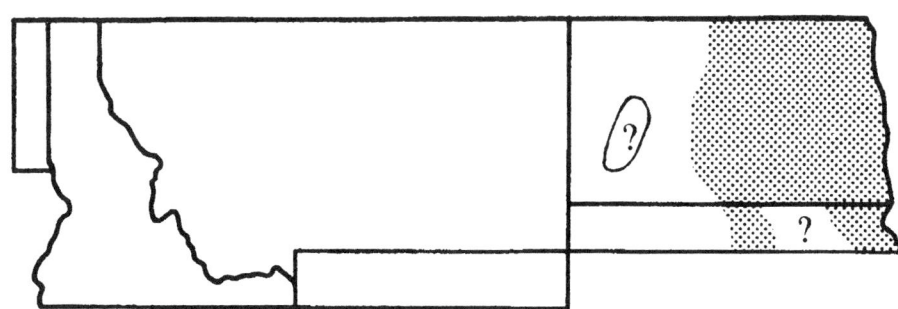

WESTERN TANAGER

Piranga ludoviciana
Thraupinae

Summer Resident

WINTERING AREA: 3

HABITAT REQUIREMENTS: Montane coniferous forests and woodlands, riparian woodlands, aspen groves (preferably mixed with conifers [Scott and Crouch, 1988b]), mountain mahogany woodland.

FEEDING: Insects, fruits, and buds taken by gleaning from foliage in trees and by hawking flying insects.

STATUS AND MANAGEMENT: Numbers are declining significantly both in Idaho and in Montana. In the West as a whole and in the North American range overall there is a slight declining trend. Generally more abundant in old-growth Douglas fir/ponderosa pine than in rotation-age forest (Hejl and Woods, 1991; Hejl et al., in preparation). Responds positively to controlled burning of ponderosa pine forest or woodland (Bock and Bock, 1983). Rare cowbird host. In spite of being widespread in western North America, the breeding biology and ecology of Western Tanagers has been little studied.

FURTHER READING: Hejl et al., 1988.

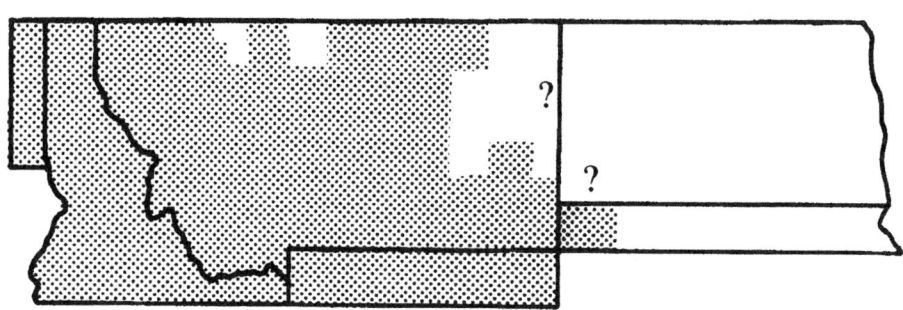

LAZULI BUNTING

Passerina amoena
Cardinalinae

Summer Resident

WINTERING AREA: 5

HABITAT REQUIREMENTS: Riparian thickets and woodlands, brush-filled canyons, aspen woodland; often found near water. Nests in shrubs or tangled undergrowth.

FEEDING: Diet composed of insects and seeds collected by foraging on the ground and by gleaning from foliage in trees and shrubs.

STATUS AND MANAGEMENT: Numbers appear to be increasing in Idaho and North Dakota but decreasing slightly in Montana. Numbers in the West overall appear to be stable with slight widespread increases reported in the most recent survey year. Numbers increased greatly this century with the creation of extensive areas of early successional habitat in the wake of logging operations and with the increase of riparian thickets in arid and semiarid regions resulting from the creation of agricultural irrigation systems. Range now appears to have contracted in the face of increased development and suburbanization in parts of the West. Riparian populations are affected negatively by livestock grazing (Mosconi and Hutto, 1982). Uncommon cowbird host.

FURTHER READING: Emlen et al., 1975; Young, 1991.

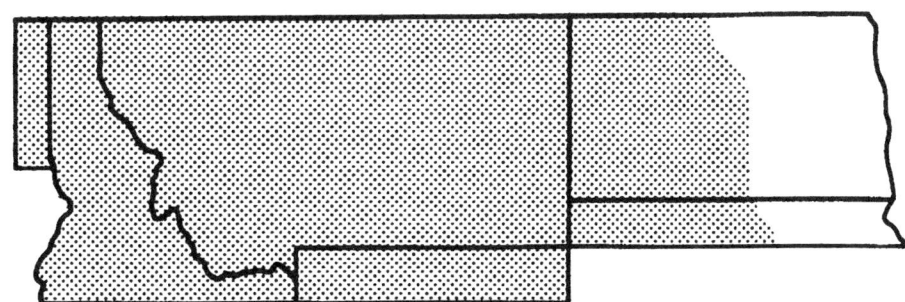

INDIGO BUNTING

Passerina cyanea
Cardinalinae

Summer Resident

WINTERING AREA: 3

HABITAT REQUIREMENTS: Open, upland and floodplain deciduous forest and woodland, including riparian thickets and second growth. Nests usually in deciduous trees or shrubs, occasionally in tangled undergrowth.

FEEDING: Consumes insects, seeds (including waste grain), berries, and other small fruits. Forages primarily by gleaning from foliage but also feeds on the ground.

STATUS AND MANAGEMENT: Numbers show significant increases in North Dakota, and a trend of increasing numbers overall in the West as Indigo Buntings have expanded their range westward. Continent-wide, however, numbers are declining with significant widespread decreases in the most recent survey years. Abundance increased this century in response to creation of favored, early successional habitats resulting from logging operations and abandonment of pastures. Frequent cowbird host.

FURTHER READING: Carey and Nolan, 1979; Emlen et al., 1975; Payne and Payne, 1990; Westneat, 1988, 1989.

DICKCISSEL

Spiza americana
Cardinalinae

Summer Resident

WINTERING AREA: 3

HABITAT REQUIREMENTS: Meadows, weedy croplands, alfalfa hayfields, ungrazed native prairie and other grasslands with scattered shrubs. Nests above ground by interweaving nest with live forbs.

FEEDING: Adults consume mostly insects, with approximately a third of diet comprised of grain, grass, and forb seeds; diet of young birds is the reverse of adult diet. Forages on the ground, gleaning items from the substrate.

STATUS AND MANAGEMENT: Populations have undergone a sustained, significant decline in North Dakota and in their entire range overall. Populations are inherently unstable with pronounced fluctuations in local numbers between years being the rule, rather than the exception. Mowing machines in alfalfa and clover fields destroy nests and nestlings. Frequent cowbird host. Dickcissels roost in large flocks following breeding.

FURTHER READING: Fretwell, 1986; Harmeson, 1974; Zimmerman, 1982.

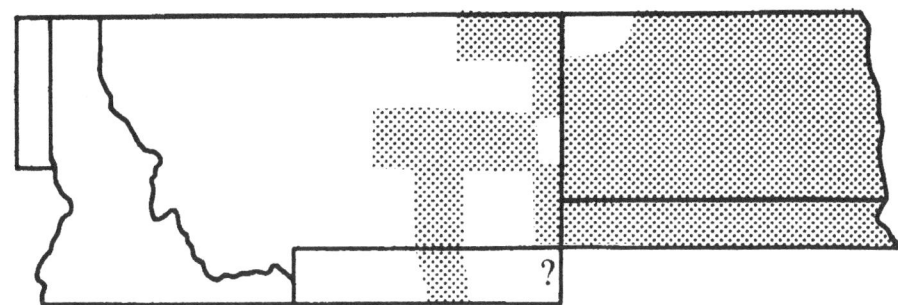

ROSE-BREASTED GROSBEAK

Pheucticus ludovicianus
Cardinalinae

Summer Resident

WINTERING AREA: 3

HABITAT REQUIREMENTS: Open, upland and floodplain deciduous forest and woodland, including riparian thickets, second growth. Nests usually in deciduous trees, less commonly in shrubs.

FEEDING: Takes insects, seeds, fruits, buds, and flowers generally by gleaning from foliage within trees and shrubs (occasionally on the ground), hovering to glean vegetation or bark, and hawking flying insects.

STATUS AND MANAGEMENT: Regional BBS data for this species are lacking, as are western data for this eastern species. For eastern North America as a whole, populations appear stable, although there are widespread significant declines in the most recent survey years. Common cowbird host.

FURTHER READING: Dunham, 1966; Holmes, 1986; Kroodsma, 1974.

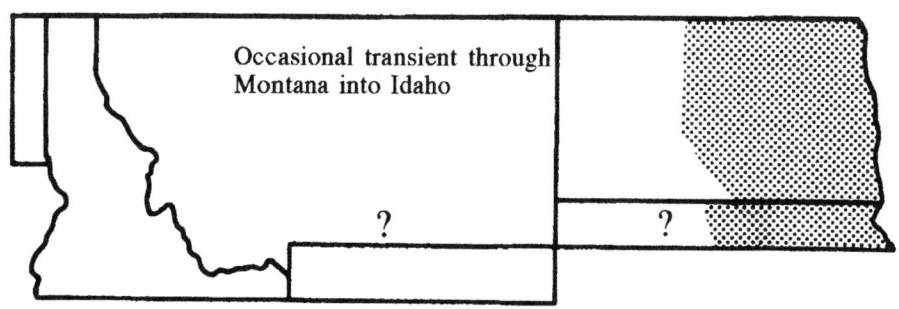

BLACK-HEADED GROSBEAK

Pheucticus melanocephalus
Cardinalinae

Summer Resident

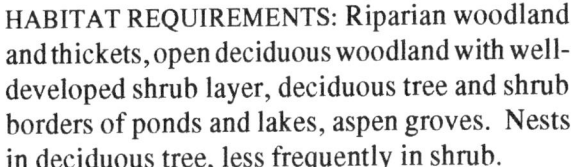

WINTERING AREA: 5

HABITAT REQUIREMENTS: Riparian woodland and thickets, open deciduous woodland with well-developed shrub layer, deciduous tree and shrub borders of ponds and lakes, aspen groves. Nests in deciduous tree, less frequently in shrub.

FEEDING: Takes insects and spiders, seeds, fruits, occasionally including buds. Forages by gleaning from foliage as it moves through trees and shrubs.

STATUS AND MANAGEMENT: Numbers are somewhat erratic in Idaho but appear to be declining rather sharply there while increasing slightly in Montana and North Dakota. In the western range overall there is a slight declining trend for this western species, although numbers have increased significantly in some portions of the range in the most recent survey year. Uncommon cowbird host.

FURTHER READING: Hill, 1988; Kroodsma, 1974.

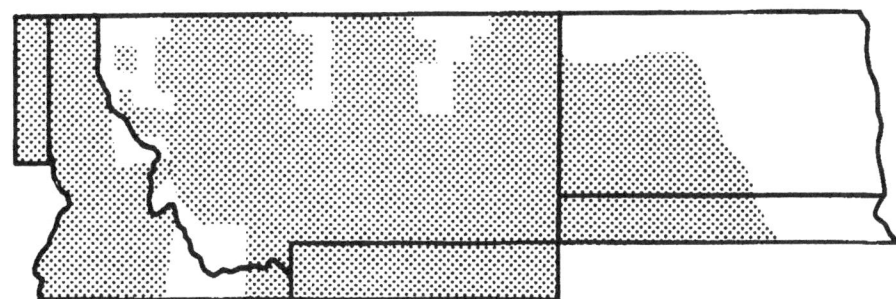

GREEN-TAILED TOWHEE

Pipilo chlorurus
Emberizinae

Summer Resident

WINTERING AREA: 5

HABITAT REQUIREMENTS: Brushy montane slopes, shrubsteppe, riparian scrub. Nests low in shrub, sometimes on the ground beneath low, dense shrub.

FEEDING: Diet of insects, grass and forb seeds, and berries is acquired primarily by scratching in leaf litter on the ground, usually beneath dense shrub-cover.

STATUS AND MANAGEMENT: Declining slightly in (southern) Idaho but significantly declining in Montana, although sample sizes are rather small. The overall trend in the range as a whole is one of relative stability. Preferred habitats contain a mixture of shrub species, almost invariably including mountain mahogany (Knopf et al., 1990). Uncommon cowbird host. Breeding biology has been little studied.

FURTHER READING: Morton, 1991.

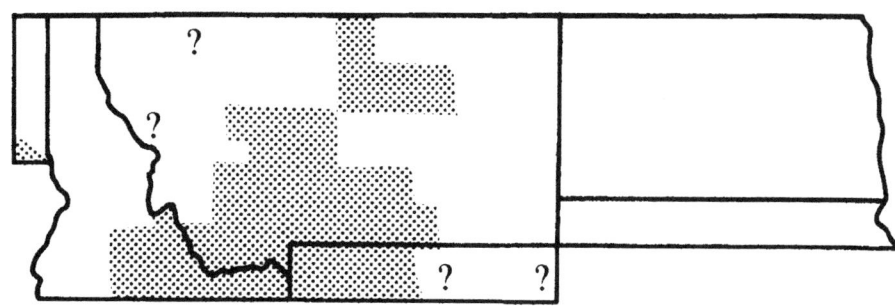

RUFOUS-SIDED TOWHEE

Pipilo erythrophthalmus
Emberizinae

Summer Resident for the most part, occasionally a Permanent Resident from westernmost Montana westward and in scattered parts of North Dakota

WINTERING AREA: 6

HABITAT REQUIREMENTS: Montane shrub-covered slopes and dense shrub undergrowth of riparian thickets, deciduous or coniferous forest edge, second-growth and open woodlands. Nests usually on the ground in slight depression beneath shrub, occasionally in low shrub.

FEEDING: Typically forages on the ground by scratching in leaf litter for insects and other terrestrial invertebrates; also gleans from foliage in shrubs and undergrowth, taking seeds of grasses and forbs as well as small fruits.

STATUS AND MANAGEMENT: Idaho populations have undergone a highly significant, steep decline, while numbers in Montana appear to be increasing and North Dakota populations appear stable. Western populations overall appear stable or slightly increasing, while numbers continent-wide show a significant declining trend with widespread significant declines in the most recent survey years. May exhibit strong site tenacity leading to sustained loss of local populations in response to habitat perturbation, whether natural (e.g., flooding) or human induced (e.g., clear-cutting) (Knopf and Sedgwick, 1987). Frequent cowbird host.

FURTHER READING: Greenlaw, 1978; Morimoto and Wasserman, 1991; Yahner, 1991.

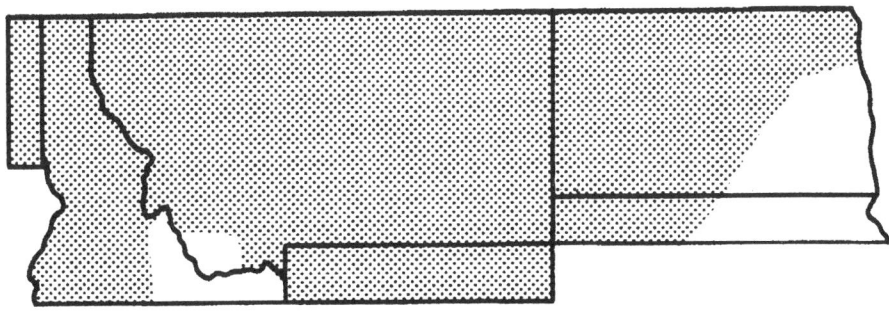

CHIPPING SPARROW

Spizella passerina
Emberizinae

Summer Resident, rarely a Winter Resident in westernmost Montana

WINTERING AREA: 5

HABITAT REQUIREMENTS: Open forests and woodlands of all types, montane scrub, partially wooded areas around human habitations. Generally nests low in trees, sometimes in tangled undergrowth, and rarely on the ground in well-concealed locale.

FEEDING: Mostly insects and spiders but also takes seeds of grasses and forbs. Forages primarily on the ground, but often gleans from foliage in trees and shrubs and occasionally hawks flying insects.

STATUS AND MANAGEMENT: Significant population declines are occurring in Idaho and Montana in contrast to a significant increasing trend in North Dakota. The overall trend for western populations is one of significant decline, but considering the species continent-wide numbers appear stable overall with widely scattered significant increases and decreases seen in the most recent survey years. Affinity for open woodland generally enables Chipping Sparrows to occupy disturbed successional habitats following logging operations. Responds positively to controlled burns in ponderosa pine woodland (Bock and Bock, 1983; Hutto, in prep.). Chipping Sparrows appear to be most abundant in the oldest rotation-age stands of mixed conifer forest (Hejl and Woods, 1991; Mannan and Meslow, 1984; Tobalske et al., 1991); increasing timber harvest at younger age-classes may be an important factor in the sparrow's general decline in the West. Frequent cowbird host.

FURTHER READING: Benkman and Pulliam, 1988; Buech, 1982; Dawson et al., 1979; Graham, 1988; Hebrard, 1978.

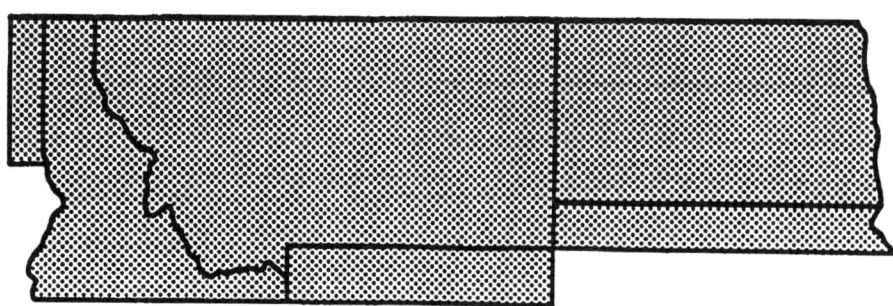

CLAY-COLORED SPARROW

Spizella pallida
Emberizinae

Summer Resident

WINTERING AREA: 5

HABITAT REQUIREMENTS: Riparian and prairie thickets, grasslands with scattered shrubs or small trees, riparian woodland, brushy woodland margins, early successional stages following logging operations or fire, shelterbelts, abandoned cultivated fields. Generally nests low in shrub, occasionally on the ground beneath shrub.

FEEDING: Takes insects and seeds of grasses and forbs, also buds and catkins in spring. Forages primarily on the ground.

STATUS AND MANAGEMENT: Numbers are declining in Montana, strongly and significantly declining in North Dakota, including significant declines in the most recent survey years. Populations are declining significantly throughout the species' western range overall, as well as throughout its continental range as a whole. The Clay-colored Sparrow is a northern Great Plains endemic whose breeding range has extended east and north since the turn of the century as logging created abundant early successional habitats. Populations in tallgrass prairie suffer increased rates of cowbird parasitism and nest predation in smaller habitat fragments (Johnson and Temple, 1986, 1990). Common cowbird host.

FURTHER READING: Buech, 1982; Knapton, 1978; Munson, 1992.

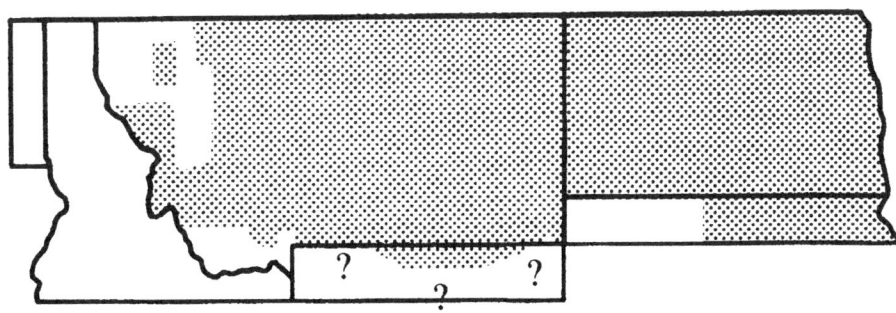

BREWER'S SPARROW

Spizella breweri
Emberizinae

Summer Resident

WINTERING AREA: 5

HABITAT REQUIREMENTS: Shrubsteppe, shortgrass prairie with scattered to abundant shrubs (especially sagebrush), montane thickets, and shrub-covered slopes. Nests in low shrub, usually sagebrush.

FEEDING: Takes insects, spiders, and seeds of grasses and forbs. Forages by gleaning items from the ground and by gleaning from shrub foliage as it moves through them.

STATUS AND MANAGEMENT: Numbers are declining steeply and significantly in Idaho and declining in Montana, but BBS data for North Dakota are lacking. Considering the western range overall, there appears to be a steady declining trend and a significant declining trend for the overall North American range, with significant declines in the central Rockies in the most recent survey years. Brewer's Sparrow is the dominant avian species associated with sagebrush habitats. Herbicidal treatment of sagebrush results in diet shift to greater proportion of seeds and a significant reduction in numbers of nesting birds in the season of spraying (Best, 1972) and complete habitat abandonment in subsequent breeding seasons (Schroeder and Sturges, 1975). Uncommon cowbird host. Many details of breeding biology are still relatively unknown.

FURTHER READING: Knopf et al., 1990; Petersen and Best, 1987; Reynolds, 1981; Rotenberry and Wiens, 1989; Short, 1984; Wiens et al., 1986, 1987, 1990.

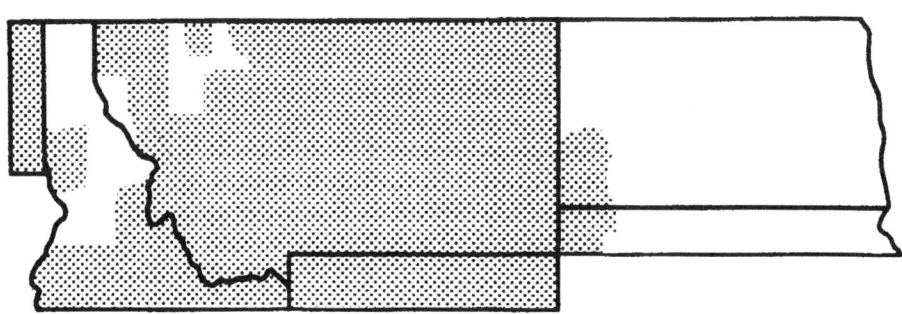

VESPER SPARROW

Pooecetes gramineus
Emberizinae

Summer Resident

WINTERING AREA: 5

HABITAT REQUIREMENTS: Generally at lower elevations in drier grasslands with scattered shrubs, prairie edge, abandoned fields, margins of shelterbelts. Nests in depression on the ground, usually well concealed.

FEEDING: Diet comprised of approximately equal proportions of insects and seeds of grasses and forbs. Forages on the ground.

STATUS AND MANAGEMENT: Numbers are increasing modestly in Idaho, declining in Montana, and significantly increasing in North Dakota. Western populations overall exhibit a slight declining trend, as is the case continent-wide, with significant recent decreases outnumbering increases in widely scattered parts of the range. Often associated with earliest seral stages of forested habitats (Hejl and Woods, 1991), and responds positively to prescribed burns in ponderosa pine forest and pine-grassland savannah (Bock and Bock, 1983). Nests in croplands are frequently destroyed by agricultural operations. Common cowbird host.

FURTHER READING: Best and Rodenhouse, 1984; George et al., 1992; Perritt and Best, 1989.

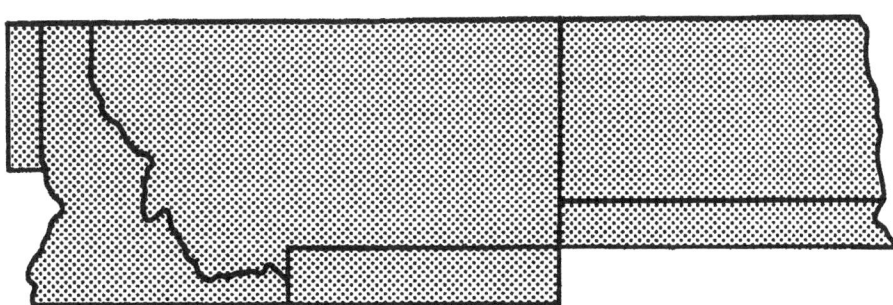

LARK SPARROW

Chondestes grammacus
Emberizinae

Summer Resident

WINTERING AREA: 5

HABITAT REQUIREMENTS: Mostly at lower elevations in prairie, often associated with farmlands or grasslands with scattered shrubs; also woodland margins and thickets adjoining grasslands or fields, agricultural areas with scattered trees and shrubs, and open woodland. Nests generally placed on the ground in slight depression, but also will nest low in shrub and sometimes in a rocky niche; known to reuse nests of other species as well as its own.

FEEDING: Diet composed largely of grass and forb seeds but also includes insects.

STATUS AND MANAGEMENT: Numbers appear to be increasing in Idaho, significantly increasing in Montana, and declining slightly in North Dakota. Western populations overall appear to be declining slightly, while continent-wide there is a significant declining trend. Lark Sparrows generally respond positively to moderate grazing pressure, depending upon habitat condition (Owens and Myres, 1973; Maher, 1979). Occasional cowbird host.

FURTHER READING: McNair, 1984; Newman, 1970.

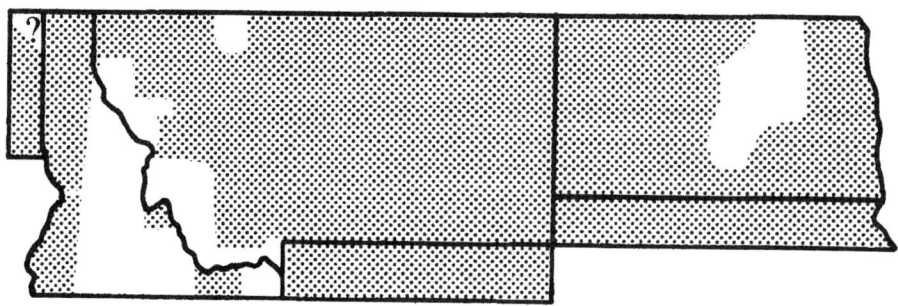

SAGE SPARROW

Amphispiza belli
Emberizinae

Summer Resident

WINTERING AREA: 5

HABITAT REQUIREMENTS: Shrubsteppe, especially sagebrush dominated. Nest usually placed in sagebrush, occasionally on ground beneath sagebrush.

FEEDING: Primarily insects and spiders but also includes some seeds. Forages on the ground and commonly gleans from foliage inside shrubs.

STATUS AND MANAGEMENT: Numbers appear to be declining in Idaho but sample sizes are very small; does not appear in BBS database for Montana. Sage Sparrows exhibit a declining trend for their range overall. Males exhibit strong site tenacity to breeding territory, even if sagebrush habitat is altered by removal of shrub cover or by burning (Wiens et al., 1986). Uncommon cowbird host.

FURTHER READING: Johnson and Marten, 1992; Petersen and Best, 1987; Reynolds, 1981; Rotenberry and Wiens, 1989; Wiens et al., 1987, 1990.

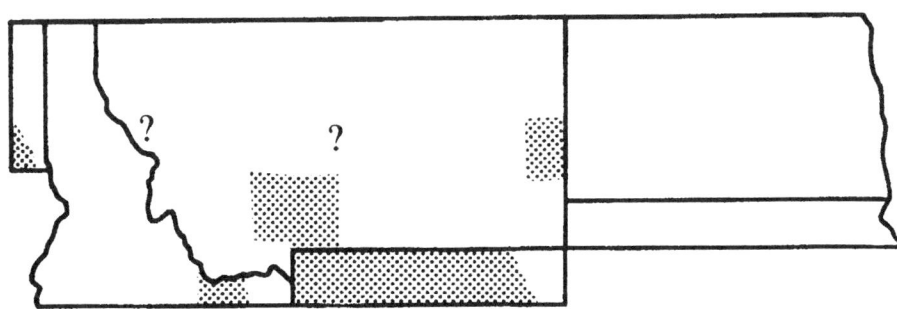

SAVANNAH SPARROW

Passerculus sandwichensis
Emberizinae

Summer Resident

WINTERING AREA: 4

HABITAT REQUIREMENTS: Moist shortgrass prairie, wet meadows from low elevations to subalpine, ungrazed or lightly grazed mixed- or tallgrass prairie, hayfields, weedy croplands, and stubble fields. Nests on the ground in slight depression, well concealed by overhanging vegetation.

FEEDING: Diet composed of insects, spiders, and other small invertebrates, as well as grass seeds. Forages on the ground, picking items from the substrate.

STATUS AND MANAGEMENT: Numbers are apparently increasing in Idaho but steadily declining in Montana and North Dakota. In the West as a whole, however, populations exhibit a significantly increasing trend, in contrast to a slight declining trend when viewed on a continent-wide basis, with widely scattered significant declines in the most recent survey year. Fragmentation of tallgrass prairie habitat produces increased rates of nest predation and brood parasitism (Johnson and Temple, 1986, 1990). Uncommon cowbird host.

FURTHER READING: Bédard and LaPointe, 1985; Bédard and Meunier, 1983; Rising, 1988; Watts, 1991; Williams, 1987.

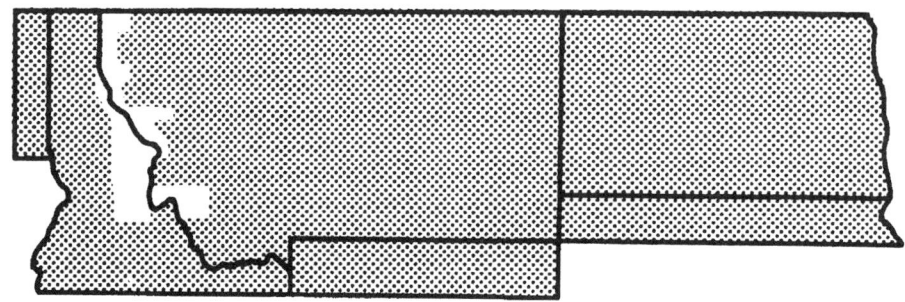

LARK BUNTING

Calamospiza melanocorys
Emberizinae

Summer Resident

WINTERING AREA: 5

HABITAT REQUIREMENTS: Prairies and other grasslands including especially those with scattered sagebrush or other low shrubs, shrubsteppe, abandoned croplands, hayfields (especially alfalfa), occasionally in stubblefields. Nests on the ground in slight depression, often at base of shrub or forb.

FEEDING: Consumes insects primarily (especially grasshoppers), also takes seeds of grasses, sedges, and forbs. Forages on the ground, picking items from the substrate; occasionally hawks flying insects.

STATUS AND MANAGEMENT: Numbers appear to be increasing in Montana but steeply and significantly declining in North Dakota, although numbers there increased significantly in the most recent survey year. Apparently declining in its western range overall, but numbers have fluctuated greatly over the period of BBS survey years; significantly declining for North American range as a whole. Destruction and fragmentation of native prairie grasslands in eastern and northeastern portion of historic range has led to contraction of range. Heavy grazing in shortgrass prairie habitats decreases suitability for nesting and can cause a decline in population density (Finch et al., 1987). Uncommon cowbird host.

FURTHER READING: Creighton, 1974; Pleszczynska, 1978; Pleszczynska and Hansell, 1980; Wunder, 1979.

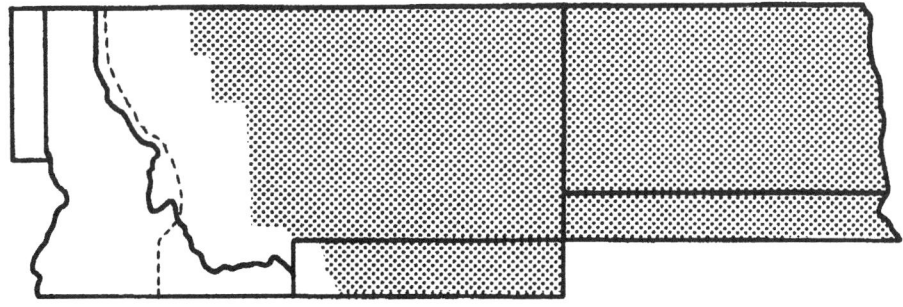

Transient westward to dotted line

BAIRD'S SPARROW

Ammodramus bairdii
Emberizinae

Summer Resident

WINTERING AREA: 5

HABITAT REQUIREMENTS: A shortgrass prairie specialist endemic to the northern Great Plains; to a lesser extent also will use mixed tallgrass/shortgrass prairie, alfalfa fields, weedy stubble fields, and retired cropland. Nests on the ground in slight depression, usually well concealed by overhanging vegetation.

FEEDING: Consumes primarily insects and spiders, also takes seeds of grasses and forbs. All foraging is done on the ground.

STATUS AND MANAGEMENT: Declining steeply and significantly in Montana, and declining at a somewhat lesser pace in North Dakota; numbers are steadily declining throughout the species' range. Listed as a Threatened Species in Canada. Apparently Baird's Sparrows do tolerate very light grazing regimens (Kantrud and Kologiski, 1982) and prefer grasslands having little or no shrub cover. Densities on native grasslands have been found to be seven times greater than densities in cultivated grasslands, and significantly more territories occurred in undisturbed grasslands than in grazed areas (Owens and Myres, 1973). Loss of suitable native prairie due to livestock grazing and agricultural conversion have caused inexorable decline in Baird's Sparrow populations throughout their range. Uncommon cowbird host.

FURTHER READING: Smith, 1987; Sousa and McDonal, 1983.

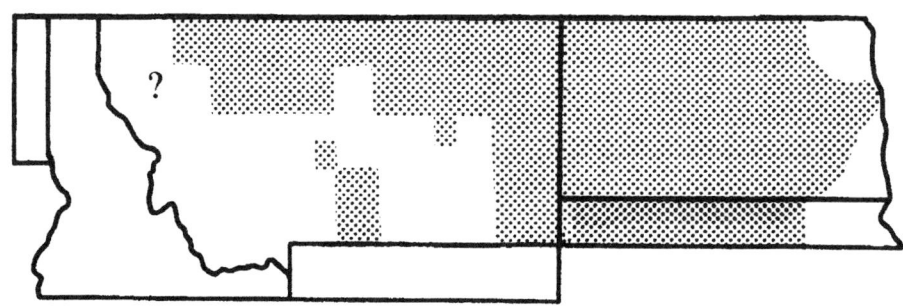

GRASSHOPPER SPARROW

Ammodramus savannarum
Emberizinae

Summer Resident

WINTERING AREA: 4

HABITAT REQUIREMENTS: Grasslands of all sorts at lower elevations: native prairie, early successional stages of abandoned agricultural fields, agricultural stubblefields, hayfields. Nests on the ground in slight depression in dense herbaceous vegetation.

FEEDING: Diet composed mostly of insects (especially grasshoppers) and other terrestrial invertebrates, also grass and forb seeds. Forages on the ground.

STATUS AND MANAGEMENT: Numbers are declining throughout the region, most sharply and significantly in North Dakota. The pattern seen in the West as a whole is one of small but steady decline, whereas continent-wide the decline has been significant and of greater magnitude, driven by relatively steeper declines in many parts of the eastern U.S. Numbers typically fluctuate locally between years. Nests in cultivated grasslands and hayfields suffer high losses from mowing. Prefers ungrazed or lightly grazed prairie, occurring only sparingly in more heavily grazed grasslands (Stewart, 1975). Conversion of native prairie to agriculture has been linked to declines in the northern Great Plains. Fragmentation of tallgrass prairie habitat produces increased rates of nest predation and brood parasitism (Johnson and Temple, 1986, 1990). Uncommon cowbird host.

FURTHER READING: Joern, 1988; Wiens, 1973.

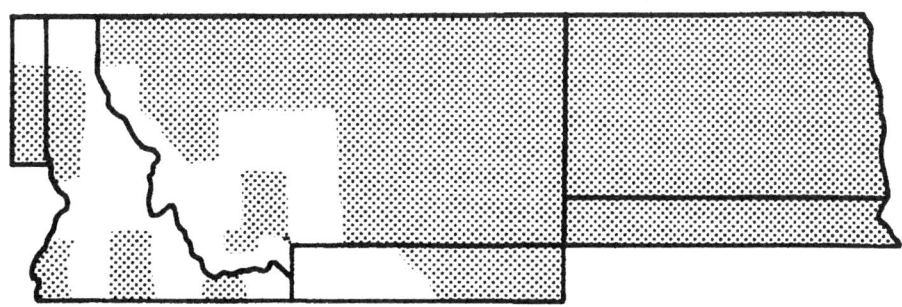

FOX SPARROW

Passerella iliaca
Emberizinae

Summer Resident, rarely a Winter Resident in north-central Montana and western Idaho

WINTERING AREA: 5

HABITAT REQUIREMENTS: Brushy forest edge, montane thickets, riparian thickets, brushy montane slopes. Nests usually placed on ground beneath shrub and occasionally low in shrub.

FEEDING: Diet composed of insects, spiders, and other terrestrial invertebrates acquired by foraging on the ground, usually scratching in leaf litter beneath shrubs or brush; also consumes seeds, berries, and buds.

STATUS AND MANAGEMENT: Sample sizes from BBS routes in the region are small but indicate relative stability in numbers in Idaho and Montana. Western populations overall exhibit a modest increasing trend; no continent-wide trend can be assessed as BBS routes are lacking in the Fox Sparrow's northern Canadian breeding range. Uncommon cowbird host.

FURTHER READING: Hejl et al., 1988; Threlfall and Blacquiere, 1982; Zink, 1986.

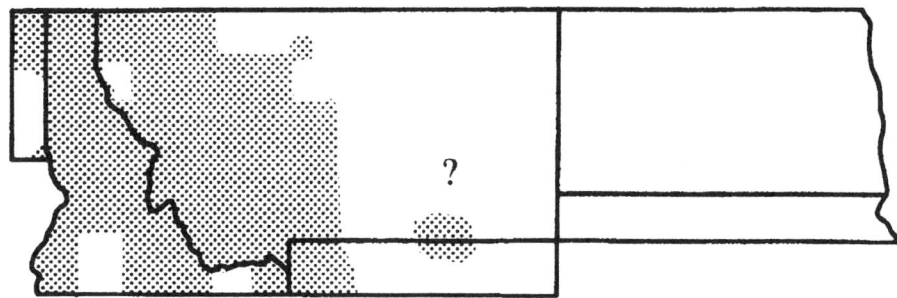

SONG SPARROW

Melospiza melodia
Emberizinae

Permanent Resident in western portion of region, primarily a Summer Resident from central Montana eastward

WINTERING AREA: 6

HABITAT REQUIREMENTS: Shrub thickets and dense brush in or near open forest or woodland, prairie thickets, riparian thickets, shrubby fringes of ponds and lakes, shelterbelts, brushy areas in vicinity of human habitations. Nests on ground beneath shrubs or in brush, as well as above ground in low shrub.

FEEDING: Consumes primarily insects, occasionally including other terrestrial invertebrates, also seeds of grasses and forbs, a few berries. Most often forages on the ground, gleaning items from the substrate but also gleans from foliage close to the ground.

STATUS AND MANAGEMENT: Populations undergoing sustained, significant decline in Idaho and North Dakota, and declining as well in Montana. Western populations overall exhibit a significant declining trend, as is also the pattern continent-wide. Along with Yellow Warblers, Song Sparrows are the most frequent targets of cowbird parasitism.

FURTHER READING: Arcese, 1989; Graham, 1988; Hochachka, 1990; Smith and Merkt, 1980; Smith et al., 1982; Stoddard et al., 1990.

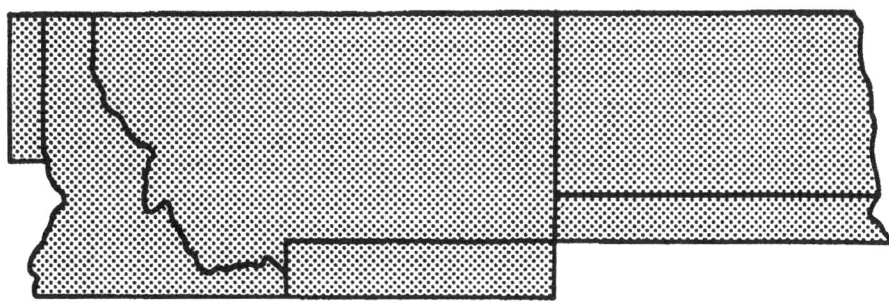

LINCOLN'S SPARROW

Melospiza lincolnii
Emberizinae

Summer Resident

WINTERING AREA: 4

HABITAT REQUIREMENTS: Montane bogs and wet meadows, willow thickets; aspen forest with shrub understory. Nests on the ground.

FEEDING: Consumes insects, spiders, and other terrestrial invertebrates, and the seeds of grasses and forbs. Forages primarily on the ground but also gleans from foliage within shrubs.

STATUS AND MANAGEMENT: Numbers are increasing in Idaho and significantly increasing in Montana. A significantly increasing trend characterizes western populations as a whole, as well as the continent-wide pattern. Riparian populations are affected negatively by livestock (Knopf et al., 1988). Rare cowbird host.

FURTHER READING: Finch and Reynolds, 1988; Raley and Anderson, 1990.

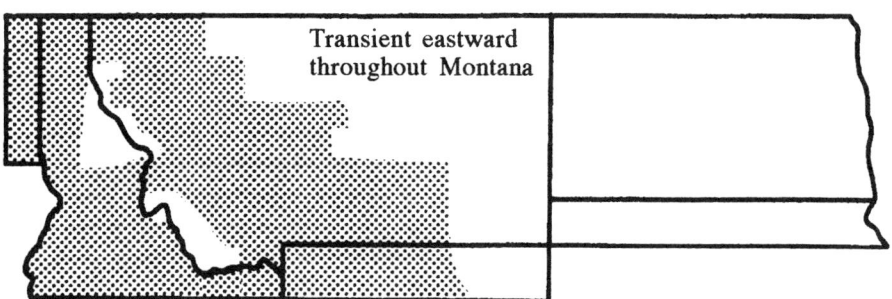
Transient eastward throughout Montana

SWAMP SPARROW

Melospiza georgiana
Emberizinae

Summer Resident

WINTERING AREA: 5

HABITAT REQUIREMENTS: Alkaline bogs and other freshwater habitats with emergent cattails or phragmites and scattered shrubs. Nests in shrub or emergent vegetation, often over shallow water, less commonly on the ground.

FEEDING: Takes terrestrial and aquatic insects, and seeds of grasses, forbs, and sedges. Forages on the ground, sometimes wades in shallow water.

STATUS AND MANAGEMENT: Populations in the region represent the northwestern edge of the species' U.S. breeding range, are highly localized, and do not appear in the regional BBS database. Continent-wide trend appears to be one of slightly increasing numbers. Common cowbird host.

FURTHER READING: Greenberg, 1988; Greenberg and Droege, 1990.

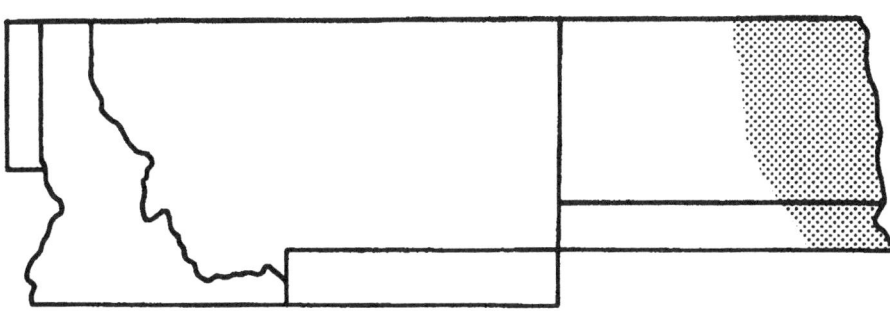

WHITE-CROWNED SPARROW

Zonotrichia leucophrys
Emberizinae

Summer Resident, occasionally Permanent Resident in westernmost portion of region

WINTERING AREA: 5

HABITAT REQUIREMENTS: Open coniferous forest with well-developed shrub understory, riparian thickets, aspen groves with shrubby understory, montane meadows with shrub thickets. Nests in shrubs or, less commonly, on ground beneath them.

FEEDING: Insects and spiders, seeds of grasses and forbs, berries and other small fruits, moss capsules and other greens. Forages mostly by gleaning from the ground but also gleans from foliage in shrubs and low trees, occasionally hawks flying insects.

STATUS AND MANAGEMENT: Populations appear to be declining in Idaho and Montana but numbers have fluctuated greatly. Both in the West as a whole and continent-wide, populations have undergone a sustained, significant declining trend. Within forested landscapes, White-crowns are associated typically with harvest units providing shrubby habitat (e.g., Moore, 1992). Riparian populations are affected negatively by livestock (Knopf et al., 1988). Uncommon cowbird host.

FURTHER READING: Benkman and Pulliam, 1988; King and Hubbard, 1981; Morton, 1992; Petrinovich and Patterson, 1983.

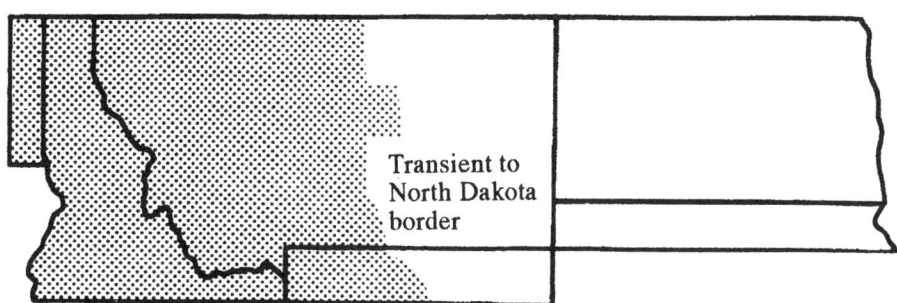

DARK-EYED JUNCO

Junco hyemalis
Emberizinae

Mostly a Permanent Resident from Montana westward and Winter Resident in North Dakota

WINTERING AREA: 6

HABITAT REQUIREMENTS: Open coniferous forest (especially ponderosa pine), edge, and woodland (especially pinyon/juniper), mixed coniferous-deciduous woodland, aspen groves, forest clearings, montane brushy fringes of ponds and bogs. Winters widely in open habitats providing shrub or tree cover. Nests on the ground in well-protected, shallow depression.

FEEDING: Forages mostly on the ground taking a wide variety of seeds and a lesser proportion of insects and spiders; also gleans from low foliage and occasionally hawks flying insects.

STATUS AND MANAGEMENT: Numbers declining sharply and significantly in Idaho, including significant declines in the most recent survey year, but numbers are apparently stable in Montana. Western populations overall appear to be declining only slightly and a similar trend is seen continent-wide. Populations respond positively to retention of logging slash, which supplies cover, foraging and nesting sites (Tobalske et al., 1991). Juncos also respond positively to controlled burns in ponderosa pine woodlands (Bock and Bock, 1983). Uncommon cowbird host.

FURTHER READING: Benkman and Pulliam, 1988; Chandler and Mulvihill, 1992; Hejl et al., 1988; Holmes and Robinson, 1988; Nolan and Ketterson, 1990; Smith and Andersen, 1982; Swanson, 1991; Wolf et al., 1991.

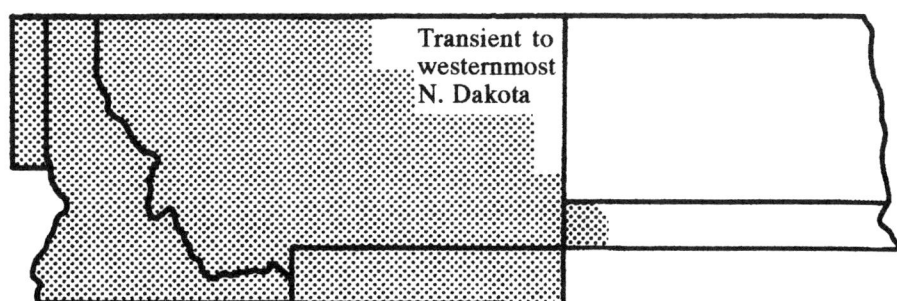

171

McCOWN'S LONGSPUR

Calcarius mccownii
Emberizinae

Summer Resident

WINTERING AREA: 5

HABITAT REQUIREMENTS: Native shortgrass prairie, moderately to heavily grazed mixed-grass prairie, small-grain stubblefields. Nests on the ground in slight depression amid sparse vegetation.

FEEDING: Diet composed mostly of insects (primarily grasshoppers, crickets, and beetles) during the breeding season, but also consumes substantial numbers of grass and forb seeds.

STATUS AND MANAGEMENT: Numbers have increased significantly in Montana, but North Dakota populations are so small and local that the species does not appear in the BBS database for the state. Although sample sizes are small, western numbers overall appear to be increasing but numbers have fluctuated greatly. In North Dakota, populations declined drastically and breeding range contracted substantially during the period 1905–1930, coincident with the extensive conversion of native prairie to agricultural land (Stewart, 1975); North Dakota birds now nest primarily in stubblefields. Uncommon cowbird host.

FURTHER READING: Greer and Anderson, 1989.

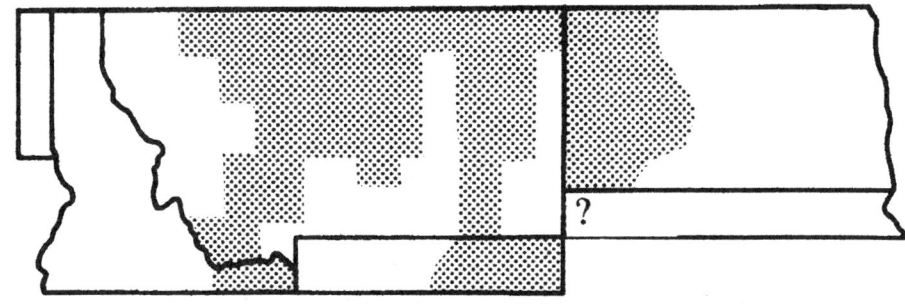

CHESTNUT-COLLARED LONGSPUR

Calcarius ornatus
Emberizinae

Summer Resident

WINTERING AREA: 5

HABITAT REQUIREMENTS: Mixed-grass and shortgrass prairie, brackish or subsaline wet meadows adjoining ponds and lakes. Nests on the ground in slight depression amid sparse vegetation.

FEEDING: Forages on the ground for insects and spiders, and the seeds of grasses, forbs, and sedges.

STATUS AND MANAGEMENT: Numbers show an increasing trend in Montana but a decreasing trend in North Dakota with a significant decrease in the most recent survey year. Western populations overall appear to be increasing slightly but numbers have fluctuated greatly. Compared to McCown's Longspur, Chestnut-collared prefers prairie grasslands that provide denser cover. Less preferred but acceptable nesting habitats include mowed hayfields and moderate to heavily grazed pastures. Numbers have been reduced greatly by conversion of native prairie habitats to agriculture (Stewart, 1975). Generally considered to be an uncommon cowbird host but may in fact be parasitized rather commonly.

FURTHER READING: Fairfield, 1968; Moriarty, 1965.

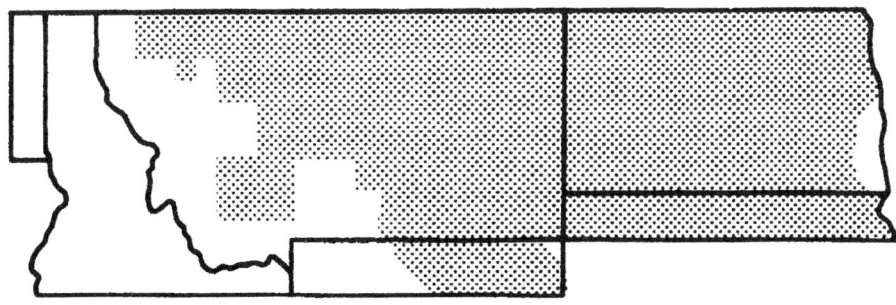

BOBOLINK

Dolichonyx oryzivorus
Icterinae

Summer Resident

WINTERING AREA: 1

HABITAT REQUIREMENTS: Tallgrass prairie, ungrazed or lightly grazed mixed-grass prairie, hayfields, wet meadows, abandoned croplands. Nests are well concealed on the ground in slight depression.

FEEDING: Insects and spiders supplemented with seeds of grasses and forbs. Forages on the ground and in low vegetation by gleaning items from the substrate and from foliage.

STATUS AND MANAGEMENT: BBS sample sizes are too small to indicate a trend for Idaho; numbers are significantly declining in Montana but are stable in North Dakota. In the West as a whole, numbers appear to be stable or even increasing, but continent-wide there is a consistent declining trend with scattered significant declines in the most recent survey years in eastern portions of the North American range. Historic range and numbers have been reduced greatly as a result of changing agricultural practices. Fragmentation of tallgrass prairie habitat can produce increased rates of nest predation and brood parasitism (Johnson and Temple, 1986, 1990). Uncommon cowbird host.

FURTHER READING: Bollinger and Gavin, 1989, 1992a, b; Gavin and Bollinger, 1988; Orians, 1985.

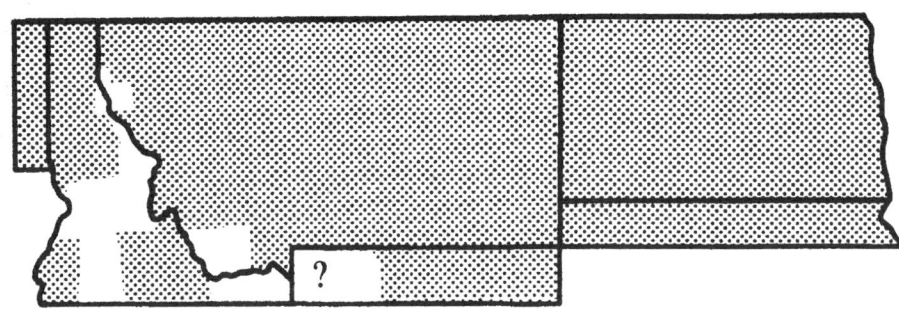

RED-WINGED BLACKBIRD

Agelaius phoeniceus
Icterinae

Summer Resident, occasionally a Winter or Permanent Resident in western portion of region

WINTERING AREA: 6

HABITAT REQUIREMENTS: Nests in wetlands of all types providing emergent vegetation: freshwater marshes, riparian habitats, wet meadows, ditches, also hayfields and croplands; forages in surrounding upland habitats. Nests in emergent vegetation, usually of reeds, cattails, or shrubs, almost invariably near or over water.

FEEDING: Takes insects and spiders, seeds of grasses and forbs, and waste grain. Forages on the ground and at water's edge, gleans from foliage in shrubs and emergent vegetation, hawks flying insects.

STATUS AND MANAGEMENT: Numbers are increasing in Idaho but declined significantly in the most recent survey year; numbers are declining significantly in Montana and apparently declining slightly in North Dakota. In the West as a whole, numbers display a small but consistent increasing trend, while continent-wide there is an overall slight declining trend, with significant overall decline in the most recent survey year. Males often form post-breeding flocks and roost together in flocks even during the summer, generally roosting in wetland areas. Extremely abundant, Red-winged Blackbirds form immense winter flocks with other blackbirds and starlings that have been the targets of occasional control efforts in agricultural and suburban areas. Frequent cowbird host.

FURTHER READING: Beletsky and Orians, 1991; Ewald and Rohwer, 1982; Moore and Dolbeer, 1989; Olson, 1991; Orians, 1980, 1985; Patterson, 1991.

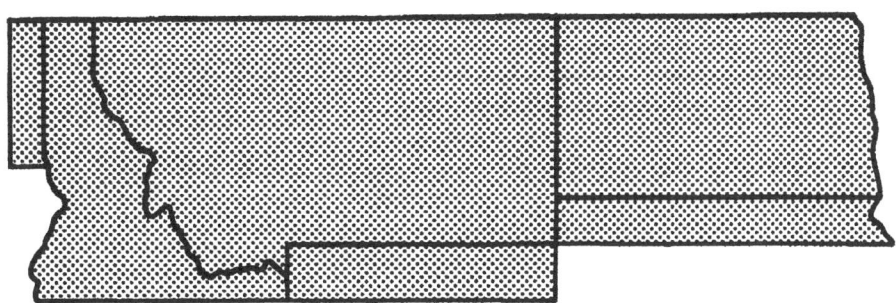

WESTERN MEADOWLARK

Sturnella neglecta
Icterinae

Icterinae

Summer Resident, occasionally a Winter or Permanent Resident at lower elevations throughout the region (especially from central Montana westward)

WINTERING AREA: 6

HABITAT REQUIREMENTS: Tallgrass and mixed-grass prairie, hayfields, wet meadows, abandoned croplands, more sparsely in shortgrass prairie and prairie with scattered sagebrush or other shrubs, occasionally in lower, montane meadows. Nests are well concealed on the ground in slight depression, often with an interwoven domed canopy.

FEEDING: Takes insects, spiders, and other terrestrial invertebrates, also includes seeds of grasses and forbs in diet. Forages on the ground, gleaning items from the substrate.

STATUS AND MANAGEMENT: Numbers are increasing modestly in Idaho, declining slightly in Montana, and stable in North Dakota. In the West as a whole, however, numbers are significantly declining in conjunction with an apparent declining trend continent-wide that includes significant declines in many parts of the range in the most recent survey year. Range has expanded to the northeast. Fragmentation of tallgrass prairie habitat can produce increased rates of nest predation and brood parasitism (Johnson and Temple, 1986, 1990). Uncommon cowbird host.

FURTHER READING: George et al., 1992; Orians, 1985; Schaeff and Picman, 1988.

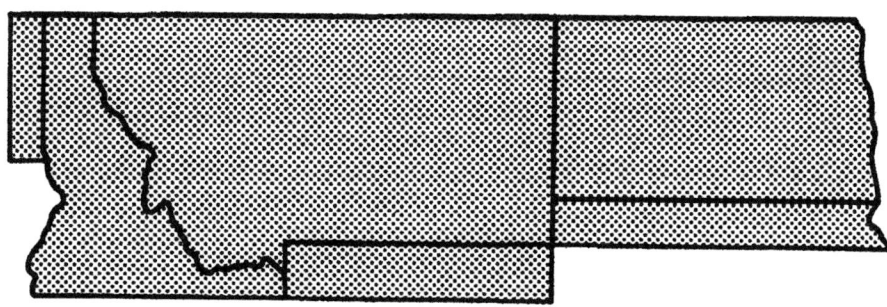

YELLOW-HEADED BLACKBIRD

Xanthocephalus xanthocephalus
Icterinae

Icterinae

Summer Resident, very rarely a Permanent Resident

WINTERING AREA: 5

HABITAT REQUIREMENTS: Nests in wetlands providing emergent vegetation, generally occupying deeper water areas than the Red-winged Blackbird; forages in surrounding upland habitats but also feeds extensively in wetlands. Nests in emergent vegetation over water.

FEEDING: Takes primarily aquatic and terrestrial insects, also spiders and seeds of grasses and forbs. Forages by gleaning items from the ground and at water's edge, hawks flying insects, and also gleans from foliage of emergent vegetation.

STATUS AND MANAGEMENT: Numbers are declining in Idaho but increasing significantly in both Montana and North Dakota, as is also the trend for the West overall and for the continent-wide range, as well. North Dakota populations increased dramatically between 1967 and 1982 (Besser, 1985); although they consume significant amounts of sunflower seeds in commercial crop fields, the economic impact of Yellow-headed Blackbirds is far less than that of the much more numerous Red-winged Blackbirds (Twedt et al., 1991). Roosts in wetlands. Rare cowbird host.

FURTHER READING: Beletsky et al., 1990; Bump, 1986; Gori, 1988; Leonard and Picman, 1986; Orians, 1980, 1985; Ortega and Cruz, 1991; Schroeder, 1982b.

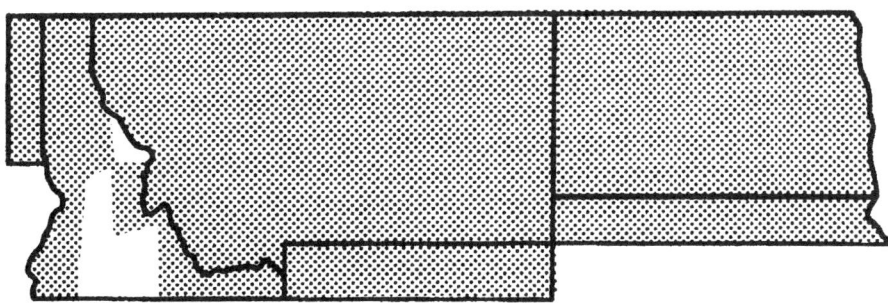

177

BREWER'S BLACKBIRD

Euphagus cyanocephalus
Icterinae

Summer Resident but rarely to occasionally a Permanent Resident in North Dakota, central and southern Montana, and westernmost Idaho

WINTERING AREA: 6

HABITAT REQUIREMENTS: Disturbed grasslands, riparian thickets and woodlands, aspen groves, agricultural lands, marshes, vicinity of human habitations. Nests colonially in trees, shrubs, emergent vegetation, and on the ground.

FEEDING: Insects, spiders, and other terrestrial as well as aquatic invertebrates are all consumed; also takes seeds of grasses and a few forbs, and small fruits. Forages primarily on the ground, occasionally wades into shallow water to forage on aquatic items, gleans from low foliage, and hawks flying insects.

STATUS AND MANAGEMENT: Numbers appear stable in Idaho, declining significantly in Montana, and increasing sharply and significantly in North Dakota. Declining significantly in the West as a whole, and declining modestly on a continent-wide basis, with a significant overall decline in the most recent survey year. Range has expanded eastward and numbers have increased greatly in response to spread of agriculture; in many areas, this species has become closely associated with agricultural and urban/suburban landscapes. The declines seen in BBS data may be more apparent than real as populations have shifted increasingly into disturbed, human-dominated landscapes and away from many of the western BBS routes. Common cowbird host.

FURTHER READING: Balph, 1975; Furrer, 1975; Horn, 1970; Orians, 1985.

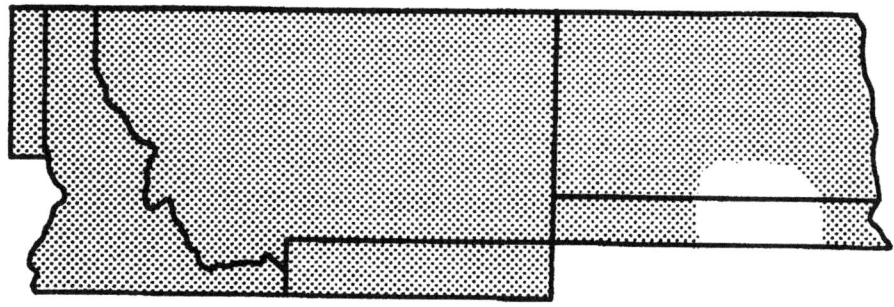

ORCHARD ORIOLE

Icterus spurius
Icterinae

Summer Resident

WINTERING AREA: 3

HABITAT REQUIREMENTS: Riparian woodlands, shelterbelts, orchards, residential areas of towns, and farms providing mature trees and shrubs. Nests in deciduous tree, less commonly in shrub.

FEEDING: Consumes insects, fruit, and tree blossoms by gleaning from foliage in trees and shrubs.

STATUS AND MANAGEMENT: Westernmost extent of breeding range is eastern Montana, where BBS sample sizes are too small to project population trends; numbers are significantly increasing in North Dakota in contrast to an overall declining trend in the western portion of the oriole's range. Populations are declining significantly when viewed continent-wide, with widespread significant declines in the most recent survey years. Common cowbird host.

FURTHER READING: Enstrom, 1992; Sealy, 1980a.

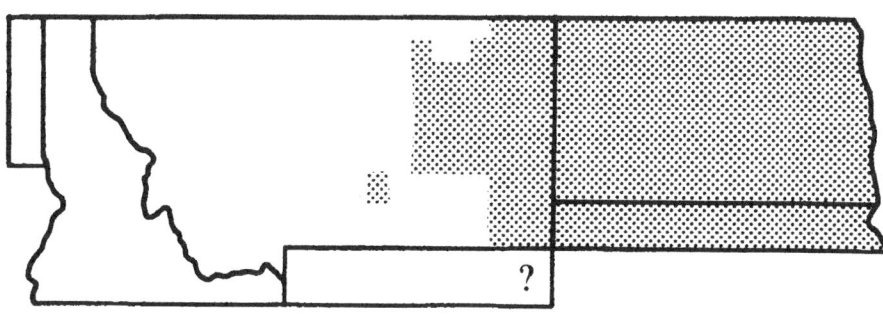

NORTHERN ORIOLE

Icterus galbula
Icterinae

Icterinae

Summer Resident

WINTERING AREA: 3

HABITAT REQUIREMENTS: Riparian and other open woodlands, open deciduous forest, shelterbelts, residential areas of towns, and farms providing mature trees and shrubs; rarely extends above lowest elevations in mountains. Nests in deciduous trees.

FEEDING: Takes insects, spiders, snails, fruit, buds, and floral nectar. Forages by gleaning from foliage in trees and by hawking flying insects.

STATUS AND MANAGEMENT: Numbers appear to be declining in Idaho, stable in Montana, and increasing in North Dakota. In the West overall, as well as continent-wide, numbers appear to be stable. Uncommon cowbird host.

FURTHER READING: Butcher, 1991; Edinger, 1988; Neudorf and Sealy, 1992; Pleasants, 1979; Rohwer and Manning, 1990; Rohwer et al., 1989; Sealy, 1980b.

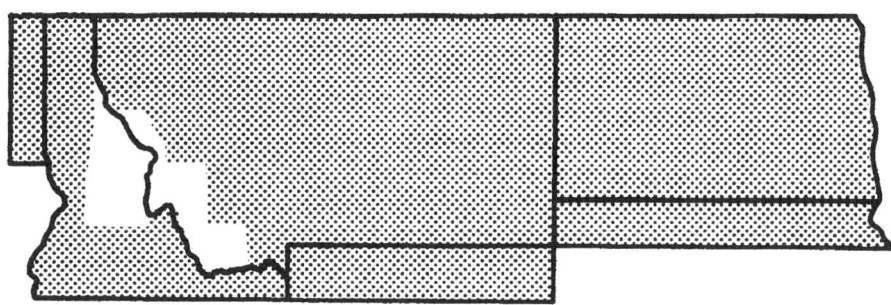

CASSIN'S FINCH

Carpodacus cassinii
Fringillidae

Summer, Permanent, or Winter Resident

WINTERING AREA: 5

HABITAT REQUIREMENTS: Drier montane coniferous forests and woodlands, especially of ponderosa pine. Nests in coniferous trees.

FEEDING: Dines primarily on seeds of conifer trees, also takes insects, buds, and berries. Forages on the ground and by gleaning from foliage in trees and shrubs.

STATUS AND MANAGEMENT: Numbers have been highly erratic in Idaho but appear to be increasing there; numbers have been more stable in Montana but appear to be declining slightly. In the West as a whole, numbers show a small but significant increasing trend. Prefers older rotation-age stands (Mannan and Meslow, 1984) and harvest units (Moore, 1992) over old growth. Cassin's Finch is a nomadic, semicolonial breeder with resultant fluctuations in local population numbers.

FURTHER READING: Hejl et al., 1988; Mewaldt and King, 1985; Samson, 1976.

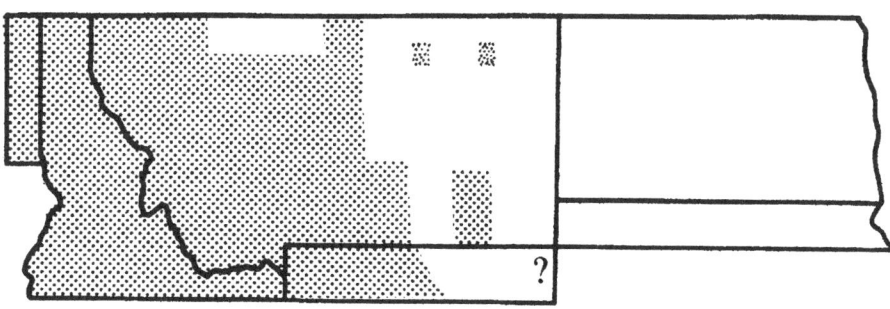

PINE SISKIN

Carduelis pinus
Fringillidae

Permanent Resident (or altitudinal migrant) from Montana westward, usually only a Summer Resident in North Dakota

WINTERING AREA: 6

HABITAT REQUIREMENTS: Coniferous and mixed coniferous-deciduous forest, only rarely in deciduous forest. Nests in coniferous tree, only rarely in deciduous tree.

FEEDING: Diet is composed primarily of seeds from coniferous and deciduous trees, grasses, and forbs, and of floral buds, nectar, and sap of trees, as well as occasional insects. Forages by gleaning from foliage in trees and by picking items from the ground.

STATUS AND MANAGEMENT: Numbers appear to be declining in Idaho where there have been marked fluctuations, stable in Montana, and declining slightly or stable in North Dakota where sample sizes are fairly small. In the West overall and continent-wide, numbers appear stable, although declines in several areas are evident in the most recent survey years. These patterns should be assessed cautiously, as populations typically fluctuate greatly between years. Pine Siskins generally prefer rotation-age forest over old growth (Hejl et al., in prep.) and respond positively to forest fragmentation (Keller and Anderson, 1992; Tobalske et al., 1991). Forms winter flocks (often mixed with goldfinches, crossbills, and juncos) that move nomadically over large areas. Uncommon cowbird host.

FURTHER READING: Hejl et al., 1988.

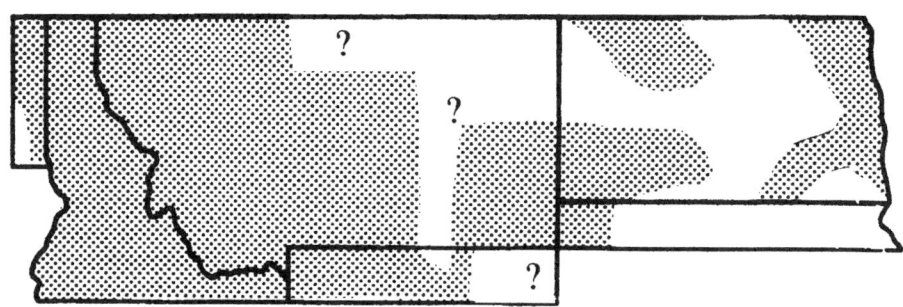

AMERICAN GOLDFINCH

Carduelis tristis
Fringillidae

Summer Resident, but occasionally a Permanent Resident from western and southwestern Montana westward

WINTERING AREA: 5

HABITAT REQUIREMENTS: Second-growth deciduous woodland, riparian woodland, abandoned fields, prairie thickets, shelterbelts, wooded residential areas. Nests in shrub or small tree, usually near water.

FEEDING: Diet composed primarily of seeds from deciduous trees, forbs (especially composites), and grasses; also takes floral buds, berries, and a few insects. Forages mostly by gleaning from foliage but also picks items from the ground.

STATUS AND MANAGEMENT: Numbers appear to be declining slightly in Idaho, increasing slightly in Montana, and increasing in North Dakota. Declining slightly in the West overall and significantly declining when viewed continent-wide, although the picture is a mixed one with significant declines in the Midwest and significant increases in the Northeast during the most recent survey year. May utilize early successional stages following woodland logging or fires. Common cowbird host, but parasite's nestlings seldom survive to fledging because of the insect-poor diet fed to goldfinch nestlings (Middleton, 1991).

FURTHER READING: Benkman and Pulliam, 1988; Skagen, 1987.

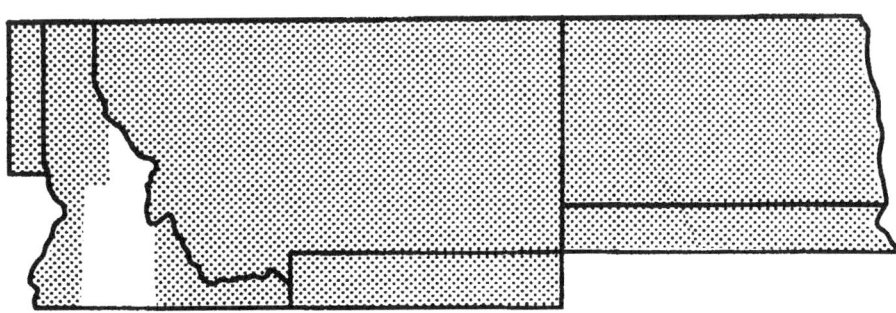

Table 3. Avian Species Occurrence in Major Habitat Types of the Northern Rockies and Great Plains*.

Species	Douglas Fir	Ponderosa Pine	Western White Pine/Larch	Lodgepole Pine	Fir/Spruce	Montane/Alpine Meadow	Aspen/Hardwoods	Montane Riparian	Plains Riparian	Shelterbelt/Woodlot	Pine/Oak Brushy Woodland Badlands/Juniper	Northern Plains Grasslands
Turkey Vulture	X	X	X	X	X	X	X	X	X	X	X	X
Osprey	X	X	X	X	X		X	X	X			
Northern Harrier						X		X	X	X	X	X
Sharp-shinned Hawk	X	X	X	X	X	X	X	X	X	X	X	X
Cooper's Hawk	X	X	X	X	X	X	X	X	X	X	X	X
Northern Goshawk	X	X	X	X	X	X	X	X	X	X	X	
Broad-winged Hawk							X	X	X	X	X	
Swainson's Hawk		X				X	X	X	X	X	X	X
Red-tailed Hawk	X	X	X	X	X	X	X	X	X	X	X	X
Ferruginous Hawk										X	X	X
Golden Eagle	X	X	X	X	X	X	X	X	X	X	X	X
American Kestrel	X	X	X	X	X	X	X	X	X	X	X	X
Merlin	X	X	X	X	X		X	X	X	X	X	X
Peregrine Falcon	X	X	X	X	X	X	X				X	
Prairie Falcon		X				X	X				X	X
Mountain Plover						X						X

Species	Killdeer	Willet	Spotted Sandpiper	Upland Sandpiper	Long-billed Curlew	Marbled Godwit	Mourning Dove	Yellow-billed Cuckoo	Black-billed Cuckoo	Flammulated Owl	Burrowing Owl	Long-eared Owl	Short-eared Owl	Common Nighthawk	Common Poorwill	Chimney Swift	Vaux's Swift	Black Swift
Northern Plains Grasslands	X	X	X	X	X	X	X		X		X		X	X	X	X		
Pine/Oak Brushy Woodland Badlands/Juniper							X		X			X		X	X			
Shelterbelt/Woodlot							X	X	X			X		X		X		
Plains Riparian	X	X	X	X	X	X	X	X	X			X	X	X		X		
Montane Riparian	X	X	X				X	X	X	X		X	X	X	X		X	X
Aspen/Hardwoods							X			X		X	X	X	X		X	X
Montane/Alpine Meadow		X		X	X	X	X				X	X	X	X			X	X
Fir/Spruce							X			X		X		X	X		X	X
Lodgepole Pine							X			X		X			X			
Western White Pine/Larch										X		X			X		X	
Ponderosa Pine							X			X		X	X	X	X		X	
Douglas Fir							X			X		X		X	X		X	X

Species	Douglas Fir	Ponderosa Pine	Western White Pine/Larch	Lodgepole Pine	Fir/Spruce	Montane/Alpine Meadow	Aspen/Hardwoods	Montane Riparian	Plains Riparian	Shelterbelt/Woodlot	Pine/Oak Brushy Woodland Badlands/Juniper	Northern Plains Grasslands
White-throated Swift	X	X	X	X	X	X	X	X				
Ruby-throated Hummingbird									X	X	X	
Black-chinned Hummingbird	X	X	X		X	X	X	X				
Calliope Hummingbird	X	X	X	X	X	X	X	X				
Broad-tailed Hummingbird	X	X	X	X	X	X	X	X				
Rufous Hummingbird	X	X	X	X	X	X	X	X				
Belted Kingfisher								X	X			
Lewis' Woodpecker	X	X	X		X		X	X				
Yellow-bellied Sapsucker								X	X	X		
Red-naped Sapsucker	X	X	X	X	X		X	X				
Williamson's Sapsucker	X	X	X	X	X		X	X				
Olive-sided Flycatcher	X	X	X	X	X		X	X	X			
Western Wood-Pewee	X	X	X	X	X		X	X	X	X		
Eastern Wood-Pewee									X	X		
Least Flycatcher						X			X	X		
Hammond's Flycatcher	X	X	X	X	X		X	X				
Dusky Flycatcher	X	X	X	X	X		X	X				

Species	Northern Plains Grasslands	Pine/Oak Brushy Woodland Badlands/Juniper	Shelterbelt/ Woodlot	Plains Riparian	Montane Riparian	Aspen/ Hardwoods	Montane/Alpine Meadow	Fir/Spruce	Lodgepole Pine	Western White Pine/ Larch	Ponderosa Pine	Douglas Fir
Willow Flycatcher	X	X	X	X	X	X	X			X	X	X
Cordilleran Flycatcher					X	X	X	X	X	X	X	X
Eastern Phoebe			X	X							X	
Say's Phoebe	X	X	X	X	X	X		X			X	
Great Crested Flycatcher		X	X	X								
Cassin's Kingbird		X	X	X							X	
Western Kingbird	X	X	X	X	X						X	
Eastern Kingbird	X	X	X	X	X							
Horned Lark	X						X					
Purple Martin			X	X	X	X	X	X	X		X	X
Tree Swallow			X	X	X	X	X	X	X	X	X	X
Violet-green Swallow				X	X	X	X	X	X	X	X	X
Northern Rough-winged Swallow	X			X	X	X		X				X
Bank Swallow	X			X	X							
Cliff Swallow	X	X		X	X	X	X	X			X	X
Barn Swallow	X	X	X	X	X	X	X	X			X	X
Brown Creeper		X	X	X	X	X		X	X	X	X	X
Rock Wren	X	X				X		X	X	X	X	X

Species	Northern Plains Grasslands	Pine/Oak Brushy Woodland Badlands/Juniper	Shelterbelt/Woodlot	Plains Riparian	Montane Riparian	Aspen/Hardwoods	Montane/Alpine Meadow	Fir/Spruce	Lodgepole Pine	Western White Pine/Larch	Ponderosa Pine	Douglas Fir
House Wren		X	X	X	X	X	X	X	X	X	X	X
Sedge Wren	X			X								
Marsh Wren	X			X	X		X					
Ruby-crowned Kinglet		X	X	X	X	X		X	X	X	X	X
Eastern Bluebird			X	X								
Western Bluebird					X	X	X	X	X	X	X	X
Mountain Bluebird					X		X	X	X	X	X	X
Townsend's Solitaire					X	X	X	X	X	X	X	X
Veery			X	X	X							
Swainson's Thrush					X	X		X	X	X	X	X
Hermit Thrush		X	X	X	X	X	X	X	X	X	X	X
American Robin		X	X	X	X	X	X	X	X	X	X	X
Gray Catbird	X	X	X	X	X	X					X	
Northern Mockingbird		X	X	X	X	X					X	
Sage Thrasher					X							
American Pipit				X	X		X					
Sprague's Pipit	X											
Cedar Waxwing		X	X	X	X	X		X				X

Species	Northern Plains Grasslands	Pine/Oak Brushy Woodland Badlands/Juniper	Shelterbelt/Woodlot	Plains Riparian	Montane Riparian	Aspen/Hardwoods	Montane/Alpine Meadow	Fir/Spruce	Lodgepole Pine	Western White Pine/Larch	Ponderosa Pine	Douglas Fir
Loggerhead Shrike	X	X	X	X	X	X	X				X	X
Bell's Vireo		X		X								
Solitary Vireo	X				X	X		X	X	X	X	X
Yellow-throated Vireo			X	X							X	X
Red-eyed Vireo		X	X	X		X					X	X
Warbling Vireo		X	X	X	X	X		X	X		X	X
Tennessee Warbler						X						
Orange-crowned Warbler		X		X	X	X		X	X		X	X
Nashville Warbler					X	X		X		X		X
Yellow Warbler			X	X	X	X	X	X	X	X	X	X
Yellow-rumped Warbler		X	X	X	X	X		X	X	X	X	X
Townsend's Warbler					X	X		X	X	X	X	X
Black-and-white Warbler		X	X	X								
American Redstart			X	X	X	X						
Ovenbird		X		X	X							
Northern Waterthrush				X	X	X				X		
MacGillivray's Warbler					X	X	X	X	X	X	X	X
Common Yellowthroat	X	X	X	X	X	X	X	X	X	X	X	X

Species	Douglas Fir	Ponderosa Pine	Western White Pine/Larch	Lodgepole Pine	Fir/Spruce	Montane/Alpine Meadow	Aspen/Hardwoods	Montane Riparian	Plains Riparian	Shelterbelt/Woodlot	Pine/Oak Brushy Woodland Badlands/Juniper	Northern Plains Grasslands
Wilson's Warbler	X	X	X	X	X	X	X	X				
Yellow-breasted Chat		X	X				X	X	X	X		X
Scarlet Tanager									X	X		
Western Tanager	X	X	X	X	X		X	X				
Lazuli Bunting	X	X			X	X	X	X	X	X	X	
Indigo Bunting		X						X	X	X	X	
Dickcissel												X
Rose-breasted Grosbeak							X	X	X	X		
Black-headed Grosbeak	X	X	X		X		X	X	X	X	X	
Green-tailed Towhee		X				X	X	X			X	
Rufous-sided Towhee	X	X	X	X	X		X	X	X	X	X	X
Chipping Sparrow	X	X	X	X	X	X	X	X	X	X	X	
Clay-colored Sparrow									X	X		X
Brewer's Sparrow						X						X
Vesper Sparrow						X					X	X
Lark Sparrow								X	X	X	X	X
Sage Sparrow						X						
Savannah Sparrow						X		X	X			X

Species	Douglas Fir	Ponderosa Pine	Western White Pine/Larch	Lodgepole Pine	Fir/Spruce	Montane/Alpine Meadow	Aspen/Hardwoods	Montane Riparian	Plains Riparian	Shelterbelt/Woodlot	Pine/Oak Brushy Woodland Badlands/Juniper	Northern Plains Grasslands
Lark Bunting											X	X
Baird's Sparrow												X
Grasshopper Sparrow											X	X
Fox Sparrow	X	X			X	X	X	X	X	X		
Song Sparrow	X	X			X	X	X	X	X	X	X	X
Lincoln's Sparrow	X			X	X	X	X	X	X			
Swamp Sparrow						X		X	X	X		
White-crowned Sparrow	X	X		X	X	X	X	X	X	X	X	X
Dark-eyed Junco	X	X	X	X	X	X	X	X	X	X	X	X
McCown's Longspur												X
Chestnut-collared Longspur												X
Bobolink									X			X
Red-winged Blackbird						X		X	X	X	X	X
Western Meadowlark						X			X	X	X	X
Yellow-headed Blackbird								X	X		X	X
Brewer's Blackbird	X	X			X	X	X	X	X		X	X
Orchard Oriole									X	X		
Northern Oriole	X	X	X		X		X	X	X	X	X	

Species	Douglas Fir	Ponderosa Pine	Western White Pine/ Larch	Lodgepole Pine	Fir/Spruce	Montane/Alpine Meadow	Aspen/ Hardwoods	Montane Riparian	Plains Riparian	Shelterbelt/ Woodlot	Pine/Oak Brushy Woodland Badlands/Juniper	Northern Plains Grasslands
Cassin's Finch	X	X	X	X	X	X	X	X				
Pine Siskin	X	X	X	X	X	X	X	X	X	X		
American Goldfinch	X	X				X	X	X	X	X	X	X

*Based in part on DeGraaf et al., 1991.

LITERATURE CITED

Alberico, J.A.R., J.M. Reed, and L.W. Oring. 1991. Nesting near a Common Tern colony increases and decreases Spotted Sandpiper nest predation. Auk 108:904-910.

Aldridge, H.D.J.N., and R.M. Brigham. 1991. Factors influencing foraging time in two aerial insectivores: the bird, *Chordeiles minor* and the bat, *Eptesicus fuscus*. Can. J. Zool. 69:62-69.

Allen, G.T. 1986. Population estimates for nesting Prairie Falcons and Golden Eagles in western North Dakota. Ph.D. Dissertation, North Dakota State Univ., Fargo.

Allen, G.T. 1987. Estimating Prairie Falcon and Golden Eagle nesting populations in North Dakota. J. Wildl. Manage. 51:739-744.

Allen, G.T., R.K. Murphy, K. Steenhof, and S.W. Platt. 1986. Late fledging dates, renesting, and large clutches of Prairie Falcons. Wilson Bull. 98:463-465.

Allen, J.N. 1980. The Ecology and Behavior of the Long-billed Curlew in Southeastern Washington. Wildl. Monogr. No. 73.

Ambuel, B., and S.A. Temple. 1982. Songbird populations in southern Wisconsin forests: 1954 and 1979. J. Field Ornithol. 53:149-158.

Andersen, D.E. 1990. Nest-defense behavior of Red-tailed Hawks. Condor 92:991-997.

Anderson, B.W., and S.A. Laymon. 1989. Creating habitat for the Yellow-billed Cuckoo (*Coccyzus americana*). In: Proceedings of the California Riparian Systems Conference. USDA Forest Serv. Gen. Tech. Rep. PSW-110, pp. 468-472.

Anderson, R.C. 1982. An evolutionary model summarizing the roles of fire, climate, and grazing animals in the origin and maintenance of grasslands. In: J.R. Estes, R.J. Tyrl, and J.N. Brunken (eds.), Grasses and Grasslands, Systematics and Ecology. Univ. Oklahoma Press, Norman, pp. 297-308.

Aney, W.C. 1984. The Effects of Patch Size on Bird Communities of Remnant Old-Growth Pine Stands in Western Montana. Thesis, Univ. Montana, Missoula.

Arad, Z., U. Midtgard, and M.H. Bernstein. 1989. Thermoregulation in Turkey Vultures: vascular anatomy, arteriovenous heat exchange, and behavior. Condor 91:505-514.

Arcese, P. 1989. Territory acquisition and loss in male Song Sparrows. Anim. Behav. 37:45-55.

Arnold, T.W., and K.F. Higgins. 1986. Effects of shrub coverages on birds of North Dakota mixed-grass prairies. Can. Field-Nat. 100:10-14.

Asay, C.E. 1987. Habitat and productivity in Cooper's Hawks. Calif. Fish Game 73:80-87.

Askins, R.A., J.F. Lynch, and R. Greenberg. 1990. Population declines in migratory birds in eastern North America. Current Ornithol. 7:1-57.

Askins, R.A., M.J. Philbrick, and D.S. Sugeno. 1987. Relationship between the regional abundance of forest and the composition of forest bird communities. Biol. Conserv. 39:129-152.

Atwood, J.L. 1992. Inferred destinations of spring migrant Common Yellowthroats based on plumage and morphology. In: J.M. Hagan, III and D.W. Johnston (eds.), Ecology and Conservation of Neotropical Migrant Landbirds. Smithsonian Inst. Press, Washington, DC, pp. 377-383.

Axelrod, D.I. 1985. Rise of the grassland biome, central North America. Biol. Rev. 51:163-201.

Baird, K.J., and J.P. Rieger. 1989. A restoration design for Least Bell's Vireo habitat in San Diego County. In: Proceedings of the California Riparian Systems Conference. USDA Forest Serv. Gen. Tech. Rep. PSW-110, pp. 462-467

Baker, M.F., R.L. Eng, J.S. Gashwiler, M.H. Schroeder, and C.E. Braun. 1976. Conservation committee report on effects of alteration of sagebrush communities on the associated avifauna. Wilson Bull. 88:165-171.

Balda, R.P. 1975. The relationship of secondary cavity nesters and snag densities in western coniferous forests. USDA For. Serv. Wildl. Habitat Tech. Bull. No. 1.

Balph, M.H. 1975. Development of young Brewer's Blackbirds. Wilson Bull. 87:207-230.

Baltosser, W.H. 1989. Nectar availability and habitat selection by hummingbirds in Guadalupe Canyon. Wilson Bull. 101:559-578.

Barbour, M.G., J.H. Burk, and W.D. Pitts. 1980. Terrestrial Plant Ecology. Benjamin/Cummings Publ., Menlo Park, CA

Barclay, R. 1977. Solitary Vireo breeding behavior. Blue Jay 35:33-37.

Barlow, J.C. 1962. Natural history of the Bell Vireo, *Vireo bellii* Audubon. Univ. Kans. Mus. Nat. Hist. Publ. 12:241-296.

Barlow, J.C. 1980. Patterns of ecological interactions among migrant and resident vireos on the wintering grounds. In: A. Keast and E.S. Morton (eds.), Migrant Birds in the Neotropics: Ecology, Behavior, Distribution, and Conservation. Smithsonian Inst. Press, Washington, DC, pp. 79-107.

Barlow, J.C., and W.B. McGillivray. 1983. Foraging and habitat relationships of the sibling species Willow Flycatcher *(Empidonax traillii)* and Alder Flycatcher *(E. alnorum)* in southern Ontario. Can. J. Zool. 61:1510-1516.

Barlow, J.C., and J.C. Rice. 1977. Aspects of the comparative behavior of Red-eyed and Philadelphia Vireos. Can. J. Zool. 55:528-542.

Barnard, P.E., B. MacWhirter, R. Simmons, G.A. Hansen, and P.C. Smith. 1987. Seasonal importance of passerine prey to Northern Harriers. Can. J. Zool. 65:1942-1946.

Beauvais, G., J.H. Enderson, and A.J. Magro. 1992. Home range, habitat use and behavior of Prairie Falcons wintering in east-central Colorado. J. Raptor Res. 26:13-18.

Beaver, D.L., and P.H. Baldwin. 1975. Ecological overlap and the problem of competition and sympatry in the Westen and Hammond's Flycatchers. Condor 77:1-13.

Bechard, M.J., R.L. Knight, D.G. Smith, and R.E. Fitzner. 1990. Nest sites of sympatric hawks (*Buteo* spp.) in Washington. J. Field Ornithol. 61:159-170.

Becker, D.M. 1985. Food habits of Richardson's Merlins in southeastern Montana. Wilson Bull. 97:226-230.

Becker, D.M., and C.H. Sieg. 1987a. Eggshell quality and organochlorine residues in eggs of Merlins, *Falco columbarius*, in southeastern Montana. Can. Field-Nat. 101:369-372.

Becker, D.M., and C.H. Sieg. 1987b. Home range and habitat utilization of breeding male Merlins, *Falco columbarius*, in southeastern Montana. Can. Field-Nat. 101:398-403.

Bédard, J., and G. LaPointe. 1985. Influence of parental age and season on Savannah Sparrow reproductive success. Condor 87:106-110.

Bédard, J., and M. Meunier. 1983. Parental care in the Savannah Sparrow. Can. J. Zool. 61:2836-2843.

Beecham, J.J., and M.N. Kochert. 1975. Breeding biology of the Golden Eagle in southwestern Idaho. Wilson Bull. 87:506-513.

Beecher, M.D., P.K. Stoddard, and P. Loesche. 1985. Recognition of parents' voices by young Cliff Swallows. Auk 102:600-605.

Beletsky, L.D., and G.H. Orians. 1991. Effects of breeding experience and familiarity on site fidelity in female Red-winged Blackbirds. Ecology 72:787-796.

Beletsky, L.D., G.H. Orians, and J.C. Wingfield. 1990. Steroid hormones in relation to territoriality, breeding density, and parental behavior in male Yellow-headed Blackbirds. Auk 107:60-68.

Belles-Isles, J.-C., and J. Picman. 1986a. House Wren nest-destroying behavior. Condor 88:190-193.

Belles-Isles, J.-C., and J. Picman. 1986b. Nesting losses and nest site preferences in House Wrens. Condor 88:483-486.

Benkman, C.W., and H.R. Pulliam. 1988. The comparative feeding rates of North American sparrows and finches. Ecology 69:1195-1199.

Bennett, S.E. 1980. Interspecific competition and the niche of the American Redstart *(Setophaga ruticilla)* in wintering and breeding communities. In: A. Keast and E.S. Morton (eds.), Migrant Birds in the Neotropics. Smithsonian Institution Press, Washington, DC, pp. 319-335.

Bergeron, D., C. Jones, D.L. Genter, and D. Sullivan. 1992. P.D. Skaar's Montana Bird Distribution, fourth edition. Special Publ. No. 2. Montana Natur. Heritage Program, Helena.

Bertin, R.I. 1977. Breeding habitats of the Wood Thrush and Veery. Auk 79:303-311.

Besser, J.F. 1985. Changes in breeding blackbird numbers in North Dakota from 1967 to 1981–82. Prairie Nat. 17:133-142.

Best, L.B. 1972. First year effects of sagebrush control on two sparrows. J. Wildl. Manage. 36:534-544.

Best, L.B., and N.L. Rodenhouse. 1984. Territory preference of Vesper Sparrows in cropland. Wilson Bull. 96:72-82.

Bider, J.R. 1968. Animal activity in uncontrolled terrestrial communities as determined by a sand transect technique. Ecol. Monogr. 38:269-308.

Bielefeldt, J., R.N. Rosenfield, and J.M. Papp. 1992. Unfounded assumptions about diet of the Cooper's Hawk. Condor 94:427-436.

Birch, T.W., and E.H. Wharton. 1982. Land-use change in Ohio, 1952–1979. USDA Forest Serv. Res. Bull. NE-70.

Birchard, G.F., and D.L. Kilgore. 1980. Conductance of water vapor in eggs of burrowing and nonburrowing birds: implications for gas exchange. Physiol. Zool. 53:284-292.

Bird, D.M. (ed.). 1983. Biology and Management of Bald Eagles and Ospreys. Harpell Press, Ste. Anne de Bellevue, Quebec.

Bjugstad, A.J., and C.F. Sorg. 1984. The value of wooded draws on the northern High Plains for hunting, furs, and woodcutting. In: D.L. Noble and R.P. Winokur (eds.), Wooded Draws: Characteristics and Values for the Northern Great Plains. South Dakota School of Mines and Technology, Agric. Publ. No. 111, Rapid City, SD, pp. 5-9.

Blair, C.L. 1978. Breeding Biology and Prey Selection of the Ferruginous Hawk in South Dakota. M.S. Thesis, Univ. South Dakota, Vermillion.

Blancher, P.J., and R.J. Robertson. 1984. Resource use by sympatric kingbirds. Condor 86:305-313.

Blancher, P.J., and R.J. Robertson. 1987. Effect of food supply on the breeding biology of Western Kingbirds. Ecology 68:723-732.

Blockstein, D.E. 1989. Crop milk and clutch size in Mourning Doves. Wilson Bull. 101:11-25.

Bock, C.E. 1970. The ecology and behavior of the Lewis Woodpecker *(Asyndesmus lewis)*. Univ. Calif. Publ. Zool. 92:1-100.

Bock, C.E., and J.H. Bock. 1983. Responses of birds and deer mice to prescribed burning in ponderosa pine. J. Wildl. Manage. 47:836-840.

Bock, C.E., and J.H. Bock. 1987. Avian habitat occupancy following fire in a Montana shrubsteppe. Prairie Nat. 19:153-158.

Bock, C.E., and D.L. Larson. 1986. Winter habitats of sapsuckers in southeastern Arizona. Condor 88:246-247.

Bock, C.E., V.A. Saab, T.D. Rich, and D.S. Dobkin. Effects of livestock grazing on Neotropical migratory landbirds in western North America. In: Workshop Proceedings of Status and Management of Neotropical Migratory Birds. In press.

Böhning-Gaese, K., M.L. Taper, and J.H. Brown. 1993. Are declines in North American insectivorous songbirds due to causes on the breeding range? Cons. Biol. 7:76-86.

Bollinger, E.K., and T.A. Gavin. 1989. The effects of site quality on breeding-site fidelity in Bobolinks. Auk 106:584-594.

Bollinger, E.K., and T.A. Gavin. 1992a. Patterns of extra-pair fertilizations in Bobolinks. Behav. Ecol. Sociobiol. 29:1-7.

Bollinger, E.K., and T.A. Gavin. 1992b. Eastern Bobolink populations: Ecology and conservation in an agricultural landscape. In: J.M. Hagan, III and D.W. Johnston (eds.), Ecology and Conservation of Neotropical Migrant Landbirds. Smithsonian Inst. Press, Washington, DC, pp. 497-506.

Bond, R.R. 1957. Ecological distribution of breeding birds in the upland forest communities of southern Wisconsin. Ecol. Monogr. 27:351-384.

Bosakowski, T. 1986. Short-eared Owl winter roosting strategies. Am. Birds 40:237-240.

Bosakowski, T., R. Kane, and D.G. Smith. 1989. Decline of the Long-eared Owl in New Jersey. Wilson Bull. 101:481-485.

Bowen, B.S., and A.D. Kruse. 1993. Effects of grazing on nesting by Upland Sandpipers in southcentral North Dakota. J. Wildl. Manage. 57:291-301.

Brawn, J.D. 1990. Interspecific competition and social behavior in Violet-green Swallows. Auk 107:606-608.

Brawn, J.D., and R.P. Balda. 1988. The influence of silvicultural activity on ponderosa pine forest bird communities in the southwestern United States. In: J.A. Jackson (ed.), Bird Conservation, Volume 3. Univ. Wisconsin Press, Madison, pp. 3-21.

Breitwisch, R., N. Gottlieb, and J. Zaias. 1989. Behavioral differences in nest visits between male and female Northern Mockingbirds. Auk 106:659-665.

Brigham, R.M. 1989. Roost and nest sites of Common Nighthawks: are gravel roofs important? Condor 91:722-724.

Brigham, R.M. 1990. Prey selection by big brown bats *(Eptesicus fuscus)* and Common Nighthawks *(Chordeiles minor)*. Am. Mid. Nat. 124:73-80.

Brigham, R.M., and M.R. Barclay. 1992. Lunar influence on foraging and nesting activity of Common Poorwills *(Phalaenoptilus nuttallii)*. Auk 109:315-320.

Briskie, J.V., and S.G. Sealy. 1989. Determination of clutch size in the Least Flycatcher. Auk 106:269-278.

Brittingham, M.C., and S.A. Temple. 1983. Have cowbirds caused forest songbirds to decline? BioScience 33:31-35.

Brooks, B.L., and S.A. Temple. 1990. Dynamics of a Loggerhead Shrike population in Minnesota. Wilson Bull. 102:441-450.

Brooks, R.P., and W.J. Davis. 1987. Habitat selection by breeding Belted Kingfishers *(Ceryle alcyon)*. Am. Mid. Nat. 117:63-70.

Brooks, R.T., and T.W. Birch. 1988. Changes in New England forests and forest owners: Implications for wildlife habitat resources and management. Trans. N.A. Wildl. Nat. Res. Conf. 53:78-87.

Brown, C.R., and M.B. Brown. 1986. Ectoparasitism as a cost of coloniality in Cliff Swallows *(Hirundo pyrrhonota)*. Ecology 67:1206-1218.

Brown, C.R., and M.B. Brown. 1988. The costs and benefits of egg destruction by conspecifics in colonial Cliff Swallows. Auk 105:737-748.

Brown, L.H., and A. Watson. 1964. The Golden Eagle in relation to its food supply. Ibis 106:78-100.

Brunton, D.H. 1988. Energy expenditure in reproductive effort of male and female Killdeer *(Charadrius vociferus)*. Auk 105:553-564.

Buchanan, J.B., C.T. Selick, L.A. Brennan, and S.G. Herman. 1988. Merlin predation on wintering Dunlins: hunting success and Dunlin escape tactics. Wilson Bull. 100:108-118.

Buech, R.R. 1982. Nesting ecology and cowbird parasitism of Clay-colored, Chipping, and Field Sparrows in a Christmas tree plantation. J. Field Ornithol. 53:363-369.

Bull, E.L. 1991. Summer roosts and roosting behavior of Vaux's Swifts in old-growth forests. Northwest. Natural. 72:78-82.

Bull, E.L., and H.D. Cooper. 1991. Vaux's Swift nests in hollow trees. Western Birds 22:85-91.

Bull, E.L., and J.E. Hohmann. 1993. The association between Vaux's Swifts and old growth forests in northeastern Oregon. Western Birds 24:38-42.

Bull, E.L., A.D. Twombly, and T.M. Quigley. 1980. Perpetuating snags in managed mixed conifer forests of the Blue Mountains, Oregon. In: R.M. DeGraff and N.G. Tilghman (eds.), Workshop Proceedings: Management of Western Forests and Grasslands for Nongame Birds. USDA For. Serv. Gen. Tech. Rep. INT-86, pp. 325-336.

Bull, E.L., A.L. Wright, and M.G. Henjum. 1989. Nesting and diet of Long-eared Owls in conifer forests, Oregon. Condor 91:908-912.

Bull, E.L., A.L. Wright, and M.G. Henjum. 1990. Nesting habitat of Flammulated Owls in Oregon. J. Raptor Res. 24:52-55.

Bump, S.R. 1986. Yellow-headed Blackbird nest defense: aggressive responses to Marsh Wrens. Condor 88:328-335.

Butcher, G.S. 1991. Mate choice in female Northern Orioles with a consideration of the role of the black male coloration in female choice. Condor 93:82-88.

Cade, T.J. 1982. The Falcons of the World. Cornell Univ. Press, Ithaca, NY.

Cade, T.J., J.H. Enderson, C.G. Thelander, and C.M. White (eds.). 1988. Peregrine Falcon Populations: Their Management and Recovery. The Peregrine Fund, Boise, Idaho.

Calder, W.A., and E.G. Jones. 1989. Implications of recapture data for migration of the Rufous Hummingbird *(Selasphorus rufus)* in the Rocky Mountains. Auk 106:488-489.

Calder, W.A., N.M. Waser, S.M. Hiebert, D.W. Inouye, and S. Miller. 1983. Site-fidelity, longevity, and population dynamics of Broad-tailed Hummingbirds: a ten-year study. Oecologia 56:359-364.

Carey, M., and V. Nolan, Jr. 1979. Population dynamics of Indigo Buntings and the evolution of avian polygyny. Evolution 33:1180-1192.

Carpenter, F.L., and M.A. Hixon. 1988. A new function for torpor: fat conservation in a wild migrant hummingbird. Condor 90:373-378.

Chandler, C.R., and R.S. Mulvihill. 1992. Effects of age, sex, and fat level on wing loading in Dark-eyed Juncos. Auk 109:235-241.

Chasko, G.G., and J.E. Gates. 1982. Avian habitat suitability along a transmission-line corridor in an oak-hickory forest region. Wildl. Monogr. 82:1-41.

Cherry, J.D. 1985. Early autumn movements and prebasic molt of Swainson's Thrushes. Wilson Bull. 97:368-370.

Clark, R.G., and R.D. Ohmart. 1985. Spread-wing posture of Turkey Vultures: single or multiple function? Condor 87:350-355.

Clark, R.J. 1975. A Field Study of the Short-eared Owl, *Asio flammeus* (Pontoppidan), in North America. Wildl. Monogr. No. 47.

Coleman, J.S., and J.D. Fraser. 1989. Habitat use and home ranges of Black and Turkey Vultures. J. Wildl. Manage. 53:782-792.

Collins, C.T. 1983. A reinterpretation of pamprodactyly in swifts: a convergent grasping mechanism in vertebrates. Auk 100:735-737.

Collins, C.T., and R.E. Landry. 1977. Artificial nest burrows for Burrowing Owls. North Am. Bird Bander 2:151-154.

Collopy, M.W. 1984. Parental care and feeding ecology of Golden Eagle nestlings. Auk 101:753-760.

Cooper, R.J., K.M. Dodge, P.J. Martinat, S.B. Donahoe, and R.C. Whitmore. 1990. Effects of diflubenzuron application on eastern deciduous forest birds. J. Wildl. Manage. 54:486-493.

Court, G.S., C.C. Gates, and D.A. Boag. 1988. Natural history of the Peregrine Falcon in the Keewatin District of the Northwest Territories. Arctic 41:17-30.

Craig, E.H., T.H. Craig, and L.R. Powers. 1988. Activity patterns and home-range use of nesting Long-eared Owls. Wilson Bull. 100:204-213.

Craig, G. 1986. Peregrine Falcon. In: R.L. Di Silvestro (ed.), Audubon Wildlife Report 1986, National Audubon Soc., NY, pp. 807-824.

Craig, R.J. 1987. Divergent prey selection in two species of waterthrushes *(Seiurus)*. Auk 104:180-187.

Craig, T.H., J.W. Connelly, E.H. Craig, and T.L. Parker. 1990. Lead concentrations in Golden and Bald Eagles. Wilson Bull. 102:130-133.

Crawford, R.D. 1977. Polygynous breeding of Short-billed Marsh Wrens. Auk 94:359-362.

Creighton, P.D. 1974. Habitat exploitation by an avian ground-foraging guild. Ph.D. Dissertation, Colorado State Univ., Fort Collins.

Crockett, A.B., and H.H. Hadow. 1975. Nest site selection by Williamson and Red-naped Sapsuckers. Condor 77:365-368.

Crockett, A.B., Jr., and P.L. Hansley. 1977. Coition, nesting, and post-fledging behavior in Williamson's Sapsucker in Colorado. Living Bird 16:7-19.

Crocoll, S.T., and J.W. Parker. 1989. The breeding biology of Broad-winged and Red-shouldered Hawks in western New York. J. Raptor Res. 23:125-139.

Cunningham, J.B., R.P. Balda, and W.S. Gaud. 1980. Selection and use of snags by secondary cavity-nesting birds of the ponderosa pine forest. USDA For. Serv. Res. Pap. RM-222.

Dale, B.C. 1984. Birds of grazed and ungrazed grasslands in Saskatchewan. Blue Jay 42:102-104.

Darley, J.A., D.M. Scott, and N.K. Taylor. 1977. Effects of age, sex and breeding success on site fidelity of Gray Catbirds. Bird-Banding 48:145-151.

Darveau, M., J.L. DesGranges, and G. Gauthier. 1992. Habitat use by three breeding insectivorous birds in declining maple forests. Condor 94:72-82.

Daubenmire, R. 1968. Ecology of fire in grasslands. Adv. Ecol. Res. 5:209-266.

Davis, C.M. 1978. A nesting study of the Brown Creeper. Living Bird 17:237-263.

Davis, W.J. 1982. Territory size in *Megaceryle alcyon* along a stream habitat. Auk 99:353-362.

Davis, W.J. 1986. Acoustic recognition in the Belted Kingfisher: cardiac response to playback vocalizations. Condor 88:505-512.

Dawson, W.R., C. Carey, C.S. Adkisson, and R.D. Ohmart. 1979. Responses of Brewer's and Chipping Sparrows to water restriction. Physiol. Zool. 52:529-541.

Debyle, N.V., and R. Winokur (eds.). 1985. Aspen: Ecology and management in the United States. USDA For. Serv. Gen. Tech. Rep. RM-119.

DeGraaf, R.M., V.E. Scott, R.H. Hamre, L. Ernst, and S.H. Anderson. 1991. Forest and Rangeland Birds of the United States. Agriculture Handbook No. 688.

DeGraff, R.M., and N.G. Tilghman (eds.), 1980. Workshop Proceedings: Management of Western Forests and Grasslands for Nongame Birds. USDA For. Serv. Gen. Tech. Rep. INT-86.

DellaSala, D.A., and D.L. Rabe. 1987. Response of Least Flycatchers *Empidonax minimus* to forest disturbances. Biol. Conserv. 41:291-299.

Derrickson, K.C. 1988. Variation in repertoire presentation in Northern Mockingbirds. Condor 90:592-606.

Dexter, R.W. 1981. Nesting success of Chimney Swifts related to age and the number of adults at the nest, and the subsequent fate of the visitors. J. Field Ornithol. 52:228-232.

Dobkin, D.S., J.A. Holmes, and B.A. Wilcox. 1986. Traditional nest-site use by White-throated Swifts. Condor 88:252-253.

Dobkin, D.S., and B.A. Wilcox. 1986. Analysis of natural forest fragments: riparian birds in the Toiyabe Mountains, Nevada. In: J. Verner, M.L. Morrison, and C.J. Ralph (eds.), Wildlife 2000: Modeling Habitat Relationships of Terrestrial Vertebrates. Univ. Wisconsin Press, Madison, pp. 293-299.

Douglas, D.C., J.T. Ratti, R.A. Black, and J.R. Alldredge. 1992. Avian habitat associations in riparian zones of Idaho's Centennial Mountains. Wilson Bull. 104:485-500.

Drilling, N.E., and C.F. Thompson. 1991. Mate switching in multibrooded House Wrens. Auk 108:60-70.

Droege, S., and J.R. Sauer. 1989. North American Breeding Bird Survey annual summary 1988. U.S. Fish Wildl. Serv., Biol. Rep. 89(13).

Dubois, K.L. 1979. An inventory of the avifauna in the Long Pines of southeastern Montana. M.S. Thesis, Montana State Univ., Bozeman.

Duebbert, H.F., and J.T. Lokemoen. 1977. Upland nesting of American Bitterns, Marsh Hawks, and Short-eared Owls. Prairie Nat. 9:33-40.

Dunham, D.W. 1966. Territorial and sexual behavior in the Rose-breasted Grosbeak. Zeit. Tierpsychol. 23:438-451.

Dunkle, S.W. 1977. Swainson's Hawks on the Laramie plains, Wyoming. Auk 94:65-71.

Eckhardt, R.C. 1976. Polygyny in the Western Wood Pewee. Condor 78:561-562.

Edinger, B.B. 1988. Extra-pair courtship and copulation attempts in Northern Orioles. Condor 90:546-554.

Edwards, T.C., Jr. 1989. The ontogeny of diet selection in fledgling Ospreys. Ecology 70:881-896.

Ehrlich, P.R., D.S. Dobkin, and D. Wheye. 1988. The Birder's Handbook: A Field Guide to the Natural History of North American Birds. Simon and Schuster, NY.

Ehrlich, P.R., D.S. Dobkin, and D. Wheye. 1992. Birds in Jeopardy: The Imperiled and Extinct Birds of the United States and Canada Including Hawaii and Puerto Rico. Stanford Univ. Press, Stanford, CA.

Ehrlich, P.R., and A.H. Ehrlich. 1990. The Population Explosion. Simon and Schuster, NY.

Emlen, S.T., J.D. Rising, and W.L. Thompson. 1975. A behavioral and morphological study of sympatry in the Indigo and Lazuli Buntings of the Great Plains. Wilson Bull. 87:145-177.

Ensign, J.T. 1983. Nest Site Selection, Productivity and Food Habits of Ferruginous Hawks in Southeastern Montana. M.S. Thesis, Montana State Univ., Bozeman.

Enstrom, D.A. 1992. Delayed plumage maturation in the Orchard Oriole *(Icterus spurius)*: tests of winter adaptation hypotheses. Behav. Ecol. Sociobiol. 30:35-42.

Erskine, A.J. 1984. Swallows foraging on the ground. Wilson Bull. 96:136-137.

Ewald, P.W. 1985. Influence of asymmetries in resource quality and age on aggression and dominance in Black-chinned Hummingbirds. Anim. Behav. 33:705-719.

Ewald, P.W., and S. Rohwer. 1982. Effects of supplemental feeding on timing of breeding, clutch size, and polygamy in Red-winged Blackbirds. J. Anim. Ecol. 51:429-450.

Faanes, C.A. 1982. Avian use of Sheyenne Lake and associated habitats in central North Dakota. U.S. Fish Wildl. Serv. Resource Publ. No. 144.

Faanes, C.A. 1983. Breeding birds of wooded draws in western North Dakota. Prairie Nat. 15:173-187.

Faanes, C.A. 1987. Breeding birds and vegetation structure in western North Dakota wooded draws. Prairie Nat. 19:209-220.

Fairfield, G.M. 1968. Chestnut-collared Longspur. In: A.C. Bent (ed.), Life Histories of North American Cardinals, Grosbeaks, Buntings, Towhees, Finches, Sparrows, and Allies, Part 3. Smithsonian Inst. Press, Washington, DC, pp. 1635-1652.

Fearnside, P.M. 1990. The rate and extent of deforestation in Brazilian Amazonia. Environ. Conserv. 17:213-226.

Feist, F.G. 1968. Breeding-bird populations on sagebrush-grassland habitat in central Montana. Aud. Field Notes 22:691-695.

Finch, D.M. 1989a. Species abundances, guild dominance patterns, and community structure of breeding riparian birds. In: R.R. Sharitz and J.W. Gibbons (eds.), Freshwater Wetlands and Wildlife. CONF-8603101, DOE Symp. Ser. No. 61. Office Sci. Tech. Inform., U.S. Dep. Energy, Oakridge, Tennessee, pp. 629-645.

Finch, D.M. 1989b. Habitat use and habitat overlap of riparian birds in three elevational zones. Ecology 70:866-880.

Finch, D.M. 1990. Effects of predation and competitor interference on nesting success of House Wrens and Tree Swallows. Condor 92:674-687.

Finch, D.M. 1991a. House Wrens adjust laying dates and clutch size in relation to annual flooding. Wilson Bull. 103:25-43.

Finch, D.M. 1991b. Population ecology, habitat requirements, and conservation of Neotropical migratory birds. USDA For. Serv. Gen. Tech. Rep. RM-205.

Finch, D.M., S.H. Anderson, and W.A. Hubert. 1987. Habitat suitability index models: Lark Bunting. U.S. Dept. Int., Fish Wildl. Serv. FWS/OBS-82/10.137.

Finch, D.M., and R.T. Reynolds. 1988. Bird response to understory variation and conifer succession in aspen forests. In: J. Emerick, S.Q. Foster, L. Hayden-Wing, J. Hodgson, J.W. Monarch, A. Smith, O. Thorne, II, and J. Todd (eds.), Issues and Technology in the Management of Impacted Western Wildlife, Symp. Proc. Thorne Ecological Institute, Boulder, Colorado, pp. 87-96.

Fitch, H.S. 1974. Observations of the food and nesting of the Broad-winged Hawk *(Buteo platypterus)* in northeastern Kansas. Condor 76:331-360.

Fitzner, R.E. 1978. The Ecology and Behavior of the Swainson's Hawk *(Buteo swainsoni)* in Southeastern Washington. Ph.D. Diss., Washington State Univ., Pullman.

Fitzner, R.E., D. Berry, L.L. Boyd, and C.A. Rieck. 1977. Nesting of Ferruginous Hawks *(Buteo regalis)* in Washington 1974-75. Condor 79:245-249.

Fitzpatrick, J.W. 1980. Wintering of North American tyrant flycatchers in the Neotropics. In: A. Keast and E.S. Morton (eds.), Migrant Birds in the Neotropics: Ecology, Behavior, Distribution, and Conservation. Smithsonian Inst. Press, Washington, DC, pp. 67-78.

Flack, J.A.D. 1976. Bird populations of aspen forest in western North America. Ornithol. Monogr. 19:1-97.

Forman, R.T.T., A.E. Galli, and C.F. Leck. 1976. Forest size and avian diversity in New Jersey woodlots with some land use implications. Oecologia 26:1-8.

Foster, M.S. 1969. Synchronized life cycles in the Orange-crowned Warbler and its mallophagan parasites. Ecology 50:315-323.

Fowler, A.C., R.L. Knight, T.L. George, and L.C. McEwen. 1991. Effects of avian predation on grasshopper populations in North Dakota grasslands. Ecology 72:1775-1781.

Frakes, R.A., and R.E. Johnson. 1982. Niche convergence in *Empidonax* flycatchers. Condor 84:286-291.

Franzreb, K.E. 1977. Bird population changes after timber harvesting of a mixed conifer forest in Arizona. USDA For. Serv. Res. Pap. RM-184.

Franzreb, K.E. 1985. Foraging ecology of Brown Creepers in a mixed-coniferous forest. J. Field Ornithol. 56:9-16.

Franzreb, K.E. 1989. Ecology and conservation of the endangered Least Bell's Vireo. U.S. Fish Wildl. Serv., Biol. Rep. 89(1).

Franzreb, K.E., and R.D. Ohmart. 1978. The effects of timber harvesting on breeding birds in a mixed-coniferous forest. Condor 80:431-441.

Fraser, J.D., and D.R. Luukkonen. 1986. The Loggerhead Shrike. In: R.L. Di Silvestro (ed.), Audubon Wildlife Report 1986, National Audubon Soc., N.Y., pp. 933-941.

Freedman, B., C. Beauchamp, I.A. McLaren, and S.I. Tingley. 1981. Forestry management practices and populations of breeding birds in a hardwood forest in Nova Scotia. Can. Field-Nat. 95:307-311.

Freemark, K.E., and H.G. Merriam. 1986. Importance of area and habitat heterogeneity to bird assemblages in temperate forest fragments. Biol. Conserv. 36:115-141.

Freer, V.M. 1979. Factors affecting site tenacity in New York Bank Swallows. Bird-Banding 50:349-357.

Fretwell, S. 1986. Distribution and abundance of the Dickcissel. Current Ornithol. 4:211-242.

Furrer, R.K. 1975. Breeding success and nest site stereotypy in a population of Brewer's Blackbirds *(Euphagus cyanocephalus)*. Oecologia 20:339-350.

Gard, N.W., and D.M. Bird. 1990. Breeding behavior of American Kestrels raising manipulated brood sizes in years of varying prey abundance. Wilson Bull. 102:605-614.

Garrison, B.A., J.M. Humphrey, and S.A. Laymon. 1987. Bank Swallow distribution and nesting ecology on the Sacramento River, California. Western Birds 18:71-76.

Garrison, B.A., R.W. Schlorff, J.M. Humphrey, S.A. Laymon, and F.J. Michny. 1989. Population trends and management of the Bank Swallow *(Riparia riparia)* on the Sacramento River, California. In: Proceedings of the California Riparian Systems Conference. USDA Forest Serv. Gen. Tech. Rep. PSW-110, pp. 267-271.

Gass, C.L. 1979. Territory regulation, tenure and migration in Rufous Hummingbirds. Can. J. Zool. 57:914-923.

Gates, J.E., and L.W. Gysel. 1978. Avian nest dispersion and fledging success in field-forest ecotones. Ecology 59:871-883.

Gaud, W.S., R.P. Balda, and J.D. Brawn. 1986. The dilemma of plots or years: a case for long-term studies. In: J. Verner, M.L. Morrison, and C.J. Ralph (eds.), Wildlife 2000: Modeling Habitat Relationships of Terrestrial Vertebrates. Univ. Wisconsin Press, Madison, pp. 223-227.

Gavin, T.A., and E.K. Bollinger. 1988. Reproductive correlates of breeding-site fidelity in Bobolinks *(Dolichonyx oryzivorus)*. Ecology 69:96-103.

Gawlik, D.E., and K.L. Bildstein. 1990. Reproductive success and nesting habitat of Loggerhead Shrikes in north-central South Carolina. Wilson Bull. 102:37-48.

George, T.L., A.C. Fowler, R.L. Knight, and L.C. McEwen. 1992. Impacts of a severe drought on grassland birds in western North Dakota. Ecol. Appl. 2:275-284.

Gibbs, J.P., and J. Faaborg. 1990. Estimating the viability of Ovenbird and Kentucky Warbler populations in forest fragments. Conserv. Biol. 4:193-196.

Gibson, D.D., and B. Kessel. 1989. Geographic variation in the Marbled Godwit and description of an Alaska subspecies. Condor 91:436-443.

Gibson, D.J., and L.C. Hulbert. 1987. Effects of fire, topography and year-to-year climate variation on species composition in tallgrass prairie. Vegetatio 72:175-185.

Gilmer, D.S., D.L. Evans, P.M. Konrad, and R.E. Stewart. 1985. Recoveries of Ferruginous Hawks banded in south-central North Dakota. J. Field Ornithol. 56:184-187.

Gilmer, D.S., P.M. Konrad, and R.E. Stewart. 1983. Nesting ecology of Red-tailed Hawks and Great Horned Owls in central North Dakota and their interactions with other large raptors. Prairie Nat. 15:133-143.

Gilmer, D.S., and R.E. Stewart. 1983. Ferruginous Hawk populations and habitat use in North Dakota. J. Wildl. Manage. 47:146-157.

Gilmer, D.S., and R.E. Stewart. 1984. Swainson's Hawk nesting ecology in North Dakota. Condor 86:12-18.

Goggans, R. 1985. Flammulated Owl Habitat Use in Northeast Oregon. M.S. Thesis, Oregon State Univ., Corvallis.

Gori, D.F. 1988. Adjustment of parental investment with mate quality by male Yellow-headed Blackbirds *(Xanthocephalus xanthocephalus)*. Auk 105:672-680.

Gradwohl, J., and R.Greenberg. 1988. Saving the Tropical Forests. Island Press, Washington, DC.

Graham, D.S. 1988. Responses of five host species to cowbird parasitism. Condor 90:588-591.

Graul, W.D. 1975. Breeding biology of the Mountain Plover. Wilson Bull. 87:6-31.

Graul, W.D., and L.E. Webster. 1976. Breeding status of the Mountain Plover. Condor 78:265-267.

Greaves, J.M. 1989. Maintaining site integrity for breeding Least Bell's Vireos. In: Proceedings of the California Riparian Systems Conference. USDA Forest Serv. Gen. Tech. Rep. PSW-110, pp. 293-298.

Green, G.A., and R.G. Anthony. 1989. Nesting success and habitat relationships of Burrowing Owls in the Columbia Basin, Oregon. Condor 91:347-354.

Greenberg, R. 1980. Demographic aspects of long-distance migration. In: A. Keast and E.S. Morton (eds.), Migrant Birds in the Neotropics: Ecology, Behavior, Distribution, and Conservation. Smithsonian Inst. Press, Washington, DC, pp. 493-504.

Greenberg, R. 1988. Water as a habitat cue for breeding Swamp and Song Sparrows. Condor 90:420-427.

Greenberg, R., and S. Droege. 1990. Adaptations to tidal marshes in breeding populations of the Swamp Sparrow. Condor 92:393-404.

Greenlaw, J.S. 1978. The relation of breeding schedule and clutch size to food supply in the Rufous-sided Towhee. Condor 80:24-33.

Greer, R.D., and S.H. Anderson. 1989. Relationships between population demography of McCown's Longspurs and habitat resources. Condor 91:609-619.

Grzybowski, J.A. 1979. Responses of Barn Swallows to eggs, young, nests, and nest sites. Condor 81:236-246.

Guinan, D.M., and S.G. Sealy. 1989. Foraging-substrate use by House Wrens nesting in natural cavities in a riparian habitat. Can. J. Zool. 67:61-67.

Haas, C.A., and S.A. Sloane. 1989. Low return rates of migratory Loggerhead Shrikes: winter mortality or low site fidelity? Wilson Bull. 101:458-460.

Hagan, J.M., III, and D.W. Johnston (eds.). 1992. Ecology and Conservation of Neotropical Migrant Landbirds. Smithsonian Inst. Press, Washington, DC.

Hagan, J.M., III, and J.R. Walters. 1990. Foraging behavior, reproductive success, and colonial nesting in Ospreys. Auk 107:506-521.

Hamerstrom, F. 1986. Harrier, Hawk of the Marshes. Smithsonian Inst. Press, Washington, DC.

Harmeson, J.P. 1974. Breeding ecology of the Dickcissel. Auk 91:348-359.

Haug, E.A., and L.W. Oliphant. 1990. Movements, activity patterns, and habitat use of Burrowing Owls in Saskatchewan. J. Wildl. Manage. 54:27-35.

Hayes, P.A., and R.J. Robertson. 1989. The impact of male parental care on female Eastern Kingbird reproductive success. Wilson Bull. 101:462-467.

Hayward, G.D., and R.E. Escano. 1989. Goshawk nest-site characteristics in western Montana and northern Idaho. Condor 91:476-479.

Hebrard, J.J. 1978. Habitat selection in two species of *Spizella*: a concurrent laboratory and field study. Auk 95:404-410.

Hejl, S.J., J. Verner, and R.P. Balda. 1988. Weather and bird populations in true fir forests of the Sierra Nevada, California. Condor 90:561-574.

Hejl, S.J., and R.E. Woods. 1991. Bird assemblages in old-growth and rotation-aged Douglas-fir/ponderosa pine stands in the northern Rocky Mountains: A preliminary assessment. In: D.M. Baumgartner and J.E. Lotan (eds.), Symposium proceedings, Interior Douglas-fir: The Species and Its Management. Wash. State Univ., Pullman, pp. 93-100.

Hendricks, B.J., and J.P. Rieger. 1989. Description of nesting habitat for Least Bell's Vireo in San Diego County. In: Proceedings of the California Riparian Systems Conference. USDA Forest Serv. Gen. Tech. Rep. PSW-110, pp. 285-292.

Hendricks, P. 1991. Repeatability of size and shape of American Pipit eggs. Can. J. Zool. 69:2624-2628.

Hendricks, P., and C.J. Norment. 1991. Effects of a severe snowstorm on subalpine and alpine populations of nesting American Pipits. J. Field Ornithol. 63:331-338.

Hendrickson, D.A., and D.M. Kubly. 1984. Desert waters: Past, present, and future. Nature Conserv. News 34:6-12.

Henny, C.J., R.A. Olson, and T.L. Fleming. 1985. Breeding chronology, molt, and measurements of *Accipiter* hawks in northeastern Oregon. J. Field Ornithol. 56:97-112.

Herlugson, C.J. 1981. Nest site selection of Mountain Bluebirds. Condor 83:252-255.

Hiebert, S.M. 1991. Seasonal differences in the response of Rufous Hummingbirds to food restriction: body mass and the use of torpor. Condor 93:526-537.

Higgins, K.F., and L. Kirsch. 1975. Some aspects of the breeding biology of the Upland Sandpiper in North Dakota. Wilson Bull. 87:96-102.

Higgins, K.F., A.D. Kruse, and J.L. Piehl. 1986. Effects of fire in the northern Great Plains. South Dak. State Univ. Ext. Circ. EC 761.

Hill, G.E. 1988. Age, plumage brightness, territory quality, and reproductive success in the Black-headed Grosbeak. Condor 90:379-388.

Hobbs, R.J., and L.F. Huenneke. 1992. Disturbance, diversity, and invasion: implications for conservation. Cons. Biol. 6:324-337.

Hochachka, W.M. 1990. Seasonal decline in reproductive performance of Song Sparrows. Ecology 71:1279-1288.

Hodorff, R.A., C.H. Sieg, and R.L. Linder. 1988. Wildlife response to stand structure of deciduous woodlands. J. Wildl. Manage. 52:667-673.

Holmes, R.T. 1986. Foraging patterns of forest birds: male-female differences. Wilson Bull. 98:196-213.

Holmes, R.T. 1990. Ecological and evolutionary impacts of bird predation on forest insects: an overview. Stud. Avian Biol. 13:6-13.

Holmes, R.T., and S.K. Robinson. 1988. Spatial patterns, foraging tactics, and diets of ground-foraging birds in a northern hardwoods forest. Wilson Bull. 100:377-394.

Holt, D.W., and J. M. Hillis. 1987. Current status and habitat associations of forest owls in western Montana. In: R.W. Nero, R.J. Clark, R.J. Knapton, and R.H. Hamre (eds.), Biology and Conservation of Northern Forest Owls. USDA Forest Serv. Gen. Tech. Rep. RM-142, pp. 281-288.

Holt, D.W., S.M. Melvin, and B. Steele. 1992. Nestling growth rates of Short-eared Owls. Wilson Bull. 104:326-333.

Holthuijzen, A.M.A. 1990. Prey delivery, caching, and retrieval rates in nesting Prairie Falcons. Condor 92:475-484.

Hopkins, R.B., J.F. Cassel, and A.J. Bjugstad. 1986. Relationships between breeding birds and vegetation in four woodland types of the Little Missouri National Grasslands. USDA For. Serv. Res. Pap. RM-270.

Horn, H.S. 1970. Social behavior of nesting Brewer's Blackbirds. Condor 72:15-23.

Houston, C.S., G.A. Fox, and R.D. Crawford. 1991. Unhatched eggs in Swainson's Hawk nests. J. Field Ornithol. 62:479-485.

Howe, F.P., and L.D. Flake. 1989. Nesting ecology of Mourning Doves in a cold desert ecosystem. Wilson Bull. 101:467-472.

Howe, M.A. 1982. Social organization in a nesting population of eastern Willets *(Catoptrophorus semipalmatus)*. Auk 99:88-102.

Howes-Jones, D. 1985. Relationships among song activity, context, and social behavior in the Warbling Vireo. Wilson Bull. 97:4-20.

Howie, R.R., and R. Ritcey. 1987 Distribution, habitat selection, and densities of Flammulated Owls in British Columbia. In: R.W. Nero, R.J. Clark, R.J. Knapton, and R.H. Hamre (eds.), Biology and Conservation of Northern Forest Owls. USDA Forest Serv. Gen. Tech. Rep. RM-142, pp. 249-254.

Hunter, R.E., J.A. Crawford, and R.E. Ambrose. 1988. Prey selection by Peregrine Falcons during the nestling stage. J. Wildl. Manage. 52:730-736.

Hunter, W.F., and P.H. Baldwin. 1962. Nesting of the Black Swift in Montana. Wilson Bull. 74:409-416.

Hurley, R.J., and E.C. Franks. 1976. Changes in the breeding ranges of two grassland birds. Auk 92:108-115.

Hutto, R.L. 1980. Winter habitat distribution of migratory landbirds in western Mexico, with special reference to small, foliage-gleaning insectivores. In: A. Keast and E.S. Morton (eds.), Migrant Birds in the Neotropics: Ecology, Behavior, Distribution, and Conservation. Smithsonian Inst. Press, Washington, DC, pp. 181-203.

Hutto, R.L. 1981a. Seasonal variation in the foraging behavior of some migratory western wood warblers. Auk 98:765-777.

Hutto, R.L. 1981b. Temporal patterns of foraging activity in some wood warblers in relation to the availability of insect prey. Behav. Ecol. Sociobiol. 9:195-198.

Hutto, R.L. 1985. Habitat selection by nonbreeding, migratory land birds. In: M.L. Cody (ed.), Habitat Selection in Birds. Academic Press, NY, pp. 455-476.

Hutto, R.L. 1986. Migratory landbirds in western Mexico: a vanishing habitat. Western Wildlands (Winter):12-16.

Hutto, R.L. 1989. The effect of habitat alteration on migratory land birds in a west Mexican tropical deciduous forest: a conservation perspective. Conserv. Biol. 3:138-148.

Hutto, R.L. 1992. Habitat distributions of migratory landbird species in western Mexico. In: J.M. Hagan, III, and D.W. Johnston (eds.), Ecology and Conservation of Neotropical Migrant Landbirds. Smithsonian Inst. Press, Washington, DC, pp. 221-239.

Inouye, D.W., W.A. Calder, and N.M. Waser. 1991. The effect of floral abundance on feeder censuses of hummingbird populations. Condor 93:279-285.

Jackson, W.M., C.S. Wood, and S. Rohwer. 1992. Age-specific plumage characters and annual molt schedules of Hermit Warblers and Townsend's Warblers. Condor 94:490-501.

James, R.D. 1976. Foraging behavior and habitat selection of three species of vireos in southern Ontario. Wilson Bull. 88: 62-75.

James, R.D. 1978. Pairing and nest site selection in Solitary and Yellow-throated Vireos with a description of a ritualized nest building display. Can. J. Zool. 56:1163-1169.

Janes, S.W. 1984a. Influences of territory composition and interspecific competition on Red-tailed Hawk reproductive success. Ecology 65:862-870.

Janes, S.W. 1984b. Fidelity to breeding territory in a population of Red-tailed Hawks. Condor 86:200-203.

Joern, A. 1988. Foraging behavior and switching by the Grasshopper Sparrow *Ammodramus savannarum* searching for multiple prey in a heterogeneous environment. Am. Mid. Nat. 119:225-234.

Johnsgard, P.A. 1979. Birds of the Great Plains. Univ. Nebraska Press, Lincoln.

Johnsgard, P.A. 1983. The Hummingbirds of North America. Smithsonian Inst. Press, Washington, DC.

Johnsgard, P.A. 1988. North American Owls. Smithsonian Inst. Press, Washington, DC.

Johnsgard, P.A. 1990. Hawks, Eagles, and Falcons of North America. Smithsonian Inst. Press, Washington, DC.

Johnson, D.R., and W.E. Melquist. 1991. Wintering distribution and dispersal of northern Idaho and eastern Washington Ospreys. J. Field Ornithol. 62:517-520.

Johnson, E.J., and L.B. Best. 1982. Factors affecting feeding and brooding of Gray Catbird nestlings. Auk 99:148-156.

Johnson, N.K. 1966. Bill size and the question of competition in allopatric and sympatric populations of Dusky and Gray Flycatchers. Syst. Zool. 15:70-87.

Johnson, N.K. 1976. Breeding distribution of Nashville and Virginia's Warblers. Auk 93:219-230.

Johnson, N.K. 1980. Character variation and evolution of sibling species in the *Empidonax difficilis-flavescens* complex (Aves: Tyrannidae). Univ. Calif. Publ. Zool. 112:1-151.

Johnson, N.K., and J.A. Marten. 1991. Evolutionary genetics of flycatchers. III. Variation in *Empidonax hammondii* (Aves: Tyrannidae). Can. J. Zool. 69:232-238.

Johnson, N.K., and J.A. Marten. 1992. Macrogeographic patterns of morphometric and genetic variation in the Sage Sparrow complex. Condor 94:1-19.

Johnson, N.K., and R.M. Zink. 1983. Speciation in sapsuckers *(Sphyrapicus)*: I. Genetic differentiation. Auk 100:871-884.

Johnson, R.G., and S.A. Temple. 1986. Assessing habitat quality for birds nesting in fragmented tallgrass prairies. In: J. Verner, M.L. Morrison, and C.J. Ralph (eds.), Wildlife 2000: Modeling Habitat Relationships of Terrestrial Vertebrates. Univ. Wisconsin Press, Madison, pp. 245-249.

Johnson, R.G., and S.A. Temple. 1990. Nest predation and brood parasitism of tallgrass prairie birds. J. Wildl. Manage. 54:106-111.

Joyce, L.A., J.E. Mitchell, and M.D. Skold (eds.). 1991. Proceedings, The Conservation Reserve—Yesterday, Today, and Tomorrow. USDA For. Serv. Gen. Tech. Rep. RM-203.

Jung, R.E. 1992. Individual variation in fruit choice by American Robins *(Turdus migratorius)*. Auk 109:98-111.

Kaiser, P.H. 1979. Upland Sandpiper nesting in southeastern South Dakota. Proc. S.D. Acad. Sci. 58:59-68.

Kantrud, H.A. 1981. Grazing intensity effects on the breeding avifauna of North Dakota native grasslands. Can. Field-Nat. 95:404-417.

Kantrud, H.A., and K.F. Higgins. 1992. Nest and nest site characteristics of some ground-nesting, non-passerine birds of northern grasslands. Prairie Nat. 24:67-84.

Kantrud, H.A., and R.L. Kologiski. 1982. Effects of soils and grazing on breeding birds of uncultivated upland grasslands of the Northern Great Plains. U.S. Dept. Int. Fish Wildl. Serv. Wildl. Res. Rep. 15.

Keast, A., and E.S. Morton (eds.). 1980. Migrant Birds in the Neotropics: Ecology, Behavior, Distribution, and Conservation. Smithsonian Inst. Press, Washington, DC.

Keast, A., and S. Saunders. 1991. Ecomorphology of the North American Ruby-crowned *(Regulus calendula)* and Golden-crowned *(R. satrapa)* Kinglets. Auk 108:880-888.

Keller, M.E., and S.H. Anderson. 1992. Avian use of habitat configurations created by forest cutting in southeastern Wyoming. Condor 94:55-65.

Kennedy, P.L., and D.R. Johnson. 1986. Prey-size selection in nesting male and female Cooper's Hawks. Wilson Bull. 98:110-115.

Kerlinger, P., and P.H. Lehrer. 1982. Owl recognition and antipredator behaviour of Sharp-shinned Hawks. Zeit. Tierpsychol. 58:163-173.

Kermott, L.H., L.S. Johnson, and M.S. Merkle. 1991. Experimental evidence for the function of mate replacement and infanticide by males in a north-temperate population of House Wrens. 1991. Condor 93:630-636.

Kilham, L. 1971. Reproductive behavior of Yellow-bellied Sapsuckers. I. Preference for nesting in *Fomes*-infected aspens and nest hole interactions with flying squirrels, raccoons, and other animals. Wilson Bull. 83:159-171.

King, J.R., and J.D. Hubbard. 1981. Comparative patterns of nestling growth in White-crowned Sparrows. Condor 83:362-369.

Kirkley, J.S., and J.A. Gessaman. 1990a. Water economy of nestling Swainson's Hawks. Condor 92:29-44.

Kirkley, J.S., and J.A. Gessaman. 1990b. Ontogeny of thermoregulation in Red-tailed Hawks and Swainson's Hawks. Wilson Bull. 102:71-83.

Kirsch, L.M., H.F. Duebbert, and A.D. Kruse. 1978. Grazing and haying effects on habitats of upland nesting birds. Trans. N.A. Wildl. Nat. Res. Conf. 43:486-497.

Kirsch, L.M., and K.F. Higgins. 1976. Upland Sandpiper nesting and management in North Dakota. Wildl. Soc. Bull. 4:16-20.

Knapton, R.W. 1978. Breeding ecology of the Clay-colored Sparrow. Living Bird 17:137-158.

Knopf, F.L. 1985. Significance of riparian vegetation to breeding birds across an altitudinal cline. In: R.R. Johnson, C.D. Ziebell, D.R. Patton, P.F. Ffolliott, and R.H. Hamre (eds.), Riparian Ecosystems and Their Management: Reconciling Conflicting Uses. USDA For. Serv. Gen. Tech. Rep. RM-120, pp. 105-111.

Knopf, F.L. 1986. Changing landscapes and the cosmopolitism of the eastern Colorado avifauna. Wildl. Soc. Bull. 14:132-142.

Knopf, F.L. 1992. Faunal mixing, faunal integrity, and the biopolitical template for diversity conservation. Trans. N.A. Wildl. & Nat. Res. Conf. 57:330-342.

Knopf, F.L., R.R. Johnson, T. Rich, F.B. Samson, and R.C. Szaro. 1988. Conservation of riparian ecosystems in the United States. Wilson Bull. 100:272-284.

Knopf, F.L., and M.L. Scott. 1990. Altered flows and created landscapes in the Platte River headwaters, 1840-1990. In: J.M. Sweeney (ed.), Management of Dynamic Ecosystems. North Cent. Sect., The Wildl. Soc., West Lafayette, Indiana, pp. 47-70.

Knopf, F.L., and J.A. Sedgwick. 1987. Latent population responses of summer birds to a catastrophic, climatological event. Condor 89:869-873.

Knopf, F.L., J.A. Sedgwick, and R.W. Cannon. 1988. Guild structure of a riparian avifauna relative to seasonal cattle grazing. J. Wildl. Manage. 52:280-290.

Knopf, F.L., J.A. Sedgwick, and D.B. Inkley. 1990. Regional correspondence among shrubsteppe bird habitats. Condor 92:45-53.

Knorr, O.A. 1961. The geographical and ecological distribution of the Black Swift in Colorado. Wilson Bull. 73:155-170.

Knowles, C.J., C.J. Stoner, and S.P. Gieb. 1982. Selective use of black-tailed prairie dog towns by Mountain Plovers. Condor 84:71-74.

Kondla, N.G. 1973. Nesting of the Black Swift at Johnston's Canyon, Alberta. Can. Field-Nat. 87:64-65.

Konrad, P.M., and D.S. Gilmer. 1984. Observations on the nesting ecology of Burrowing Owls in central North Dakota. Prairie Nat. 16:129-130.

Kroodsma, D.E. 1973. Coexistence of Bewick's Wrens and House Wrens in Oregon. Auk 90:341-352.

Kroodsma, D.E. 1989. Two North American song populations of the Marsh Wren reach distributional limits in the central Great Plains. Condor 91:332-340.

Kroodsma, R.L. 1973. Breeding bird populations of riverine forests in eastern Montana. Prairie Nat. 5:40-48.

Kroodsma, R.L. 1974. Species-recognition behavior of territorial male Rose-breasted and Black-headed Grosbeaks *(Pheucticus)*. Auk 91:54-64.

Krueger, K. 1986. Feeding relationships among bison, pronghorn, and prairie dogs: an experimental study. Ecology 67:760-770.

Kus, B.E., and K.L. Miner. 1989. Use of non-riparian habitats by Least Bell's Vireos. In: Proceedings of the California Riparian Systems Conference. USDA Forest Serv. Gen. Tech. Rep. PSW-110, pp. 299-303.

Lawrence, L. de K. 1967. A Comparative Life-History Study of Four Species of Woodpeckers. Ornithol. Monogr. No. 5.

Laymon, S.A. 1987. Brown-headed Cowbirds in California: historical perspectives and management opportunities in riparian habitats. Western Birds 18:63-70.

Laymon, S.A., and M.D. Halterman. 1987. Can the western subspecies of the Yellow-billed Cuckoo be saved from extinction? Western Birds 18:19-25.

Laymon, S.A., and M.D. Halterman. 1989. A proposed habitat management plan for Yellow-billed Cuckoos in California. In: Proceedings of the California Riparian Systems Conference. USDA Forest Serv. Gen. Tech. Rep. PSW-110, pp. 272-277.

Leck, C.F., and F.L. Cantor. 1979. Seasonality, clutch size, and hatching success in the Cedar Waxwing. Auk 96:196-198.

Lederer, J.R. 1977. Winter feeding territories in the Townsend's Solitaire. Bird-Banding 48:11-18.

Leedy, R.R. 1972. The status of Prairie Falcons in western Montana: special emphasis on possible effects of chlorinated hydrocarbon insecticides. M.S. Thesis, Univ. Montana, Missoula.

Lemon, R.E., D.M. Weary, and K.J. Norris. 1992. Male morphology and behavior correlate with reproductive success in the American Redstart *(Setophaga ruticilla)*. Behav. Ecol. Sociobiol. 29:399-404.

Leonard, M.L., and J. Picman. 1986. Why are nesting Marsh Wrens and Yellow-headed Blackbirds spatially segregated? Auk 103:135-140.

Leonard, M.L., and J. Picman. 1987. Nesting mortality and habitat selection by Marsh Wrens. Auk 104:491-495.

Leopold, A.S., R.J. Gutierrez, and M.T. Bronson. 1981. North American Game Birds and Mammals. Charles Scribner's Sons, NY.

Lewis, D.M. 1972. Importance of face mask in sexual recognition and territorial behavior in the Yellowthroat. Jack-Pine Warbler 50:98-109.

Li, P., and T.E. Martin. 1991. Nest-site selection and nesting success of cavity-nesting birds in high elevation forest drainages. Auk 108:405-418.

Linkhart, B.D., and R.T. Reynolds. 1987. Brood division and postnesting behavior of Flammulated Owls. Wilson Bull. 99:240-243.

Logan, C.A. 1991. Mate switching and mate choice in female Northern Mockingbirds: facultative monogamy. Wilson Bull. 103:277-281.

Lokemoen, J.T., and H.F. Duebbert. 1976. Ferruginous Hawk nesting ecology and raptor populations in northern South Dakota. Condor 78:464-470.

Lombardo, M.P. 1991. Sexual differences in parental effort during the nestling period in Tree Swallows *(Tachycineta bicolor)*. Auk 108:393-404.

Losito, M.P., R.E. Mirarchi, and G.A. Baldassarre. 1990. Summertime activity budgets of hatching-year Mourning Doves. Auk 107:18-24.

Lunk, W.A. 1962. The Rough-winged Swallow, *Stelgidopteryx serripennis*: a study based on its breeding biology in Michigan. Nuttal Ornith. Club Publ. No. 4, Cambridge, MA.

Lynch, J.F., and D.F. Whigham. 1984. Effects of forest fragmentation on breeding bird communities in Maryland, USA. Biol. Conserv. 28:287-324.

Lyon, L.J., and J.M. Marzluff. 1985. In: Fire's effects on wildlife habitat—symposium proceedings. USDA For. Serv. Gen. Tech. Rep. INT-186, pp. 16-22.

MacCracken, J.G., D.W. Uresk, and R.M. Hansen. 1985. Vegetation and soils of Burrowing Owl nest sites in Conata Basin, South Dakota. Condor 87:152-154.

Mack, R.N., and J.N. Thompson. 1982. Evolution in steppe with few large, hooved mammals. Am. Nat. 119:757-773.

MacKenzie, D.I., and S.G. Sealy. 1981. Nest site selection in Eastern and Western Kingbirds: a multivariate approach. Condor 83:310-321.

Maher, W.J. 1979. Nestling diets of prairie passerine birds at Matador, Saskatchewan, Canada. Ibis 121:437-452.

Mannan, R.W. 1980. Assemblages of bird species in western coniferous old-growth forests. In: R.M. DeGraff and N.G. Tilghman (eds.), Workshop Proceedings: Management of Western Forests and Grasslands for Nongame Birds. USDA For. Serv. Gen. Tech. Rep. INT-86, pp. 357-368.

Mannan, R.W., and E.C. Meslow. 1984. Bird populations and vegetation characteristics in managed and old-growth forests, northeastern Oregon. J. Wildl. Manage. 48:1219-1238.

Manuwal, D.A. 1970. Notes on the territoriality of Hammond's Flycatcher in western Montana. Condor 72:364-365.

Manuwal, D.A., and M.H. Huff. 1987. Spring and winter bird populations in a Douglas-fir forest sere. J. Wildl. Manage. 51:586-595.

Marks, J.S. 1986. Nest site characteristics and reproductive success of Long-eared Owls in southwestern Idaho. Wilson Bull. 98:547-560.

Marshall, J.T. 1988. Birds lost from a giant sequoia forest during fifty years. Condor 90:359-372.

Marti, C.D. 1974. Feeding ecology of four sympatric owls. Condor 76:45-61.

Marti, C.D. 1976. A review of prey selection by the Long-eared Owl. Condor 78:331-336.

Martin, J.W. 1987. Behavior and habitat use of breeding Northern Harriers in southwestern Idaho. J. Raptor Res. 21:57-66.

Martin, T.E. 1980. Diversity and abundance of spring migratory birds using habitat islands on the Great Plains. Condor 82:430-439.

Martin, T.E. 1988. Habitat and area effects on forest bird assemblages: is nest predation an influence? Ecology 69:74-84.

Martin, T.E. 1992. Breeding productivity considerations: What are the appropriate habitat features for management? In: J.M. Hagan, III, and D.W. Johnston (eds.), Ecology and Conservation of Neotropical Migrant Landbirds. Smithsonian Inst. Press, Washington, DC, pp. 455-473.

Martin, T.E., and J.J. Roper. 1988. Nest predation and nest-site selection of a western population of the Hermit Thrush. Condor 90:51-57.

Marvil, R.E., and A. Cruz. 1989. Impact of Brown-headed Cowbird parasitism on the reproductive success of the Solitary Vireo. Auk 106:476-480.

Matchett, M.R., and B.W. O'Gara. 1987. Methods of controlling Golden Eagle depredation on domestic sheep in southwestern Montana. J. Raptor. Res. 21:85-94.

Matray, P.F. 1974. Broad-winged Hawk nesting and ecology. Auk 91:307-324.

Maurer, B.A., and R.C. Whitmore. 1981. Foraging of five bird species in two forests with different vegetation structure. Wilson Bull. 93:478-490.

McCallum, D.A., and F.R. Gehlbach. 1988. Nest-site preferences of Flammulated Owls in western New Mexico. Condor 90:653-661.

McClelland, B.R., and S.S. Frissell. 1975. Identifying forest snags useful for hole-nesting birds. J. For. 73:414-417.

McClelland, B.R., S.S. Frissell, W.C. Fischer, and C.H. Halvorson. 1979. Habitat management for hole-nesting birds in forests of western larch and Douglas-fir. J. For. 77:480-483.

McEwen, L.C., T.L. George, and B.E. Petersen. 1990. Response of wildlife to current range grasshopper IPM practices and to former control methods. In: Environmental Contaminants and Their Effects on Biota of the Northern Great Plains. N. Dak. Chap. Wildl. Soc., Bismarck, ND.

McKitrick, M.C. 1990. Genetic evidence for multiple parentage in Eastern Kingbirds (Tyrannus tyrannus). Behav. Ecol. Sociobiol. 26:149-155.

McNair, D.B. 1984. Reuse of other species nests by Lark Sparrows. Southwest. Nat. 29:506-509.

McPherson, J.M. 1987. A field study of winter fruit preferences of Cedar Waxwings. Condor 89:293-306.

Medin, D.E. 1985. Breeding bird responses to diameter-cut logging in west-central Idaho. USDA For. Serv. Res. Pap. INT-355.

Medin, D.E., and G.D. Booth. 1989. Responses of birds and small mammals to single-tree selection logging in Idaho. USDA For. Serv. Res. Pap. INT-408.

Medin, D.E., and W.P. Clary. 1990. Bird and small mammal populations in a grazed and ungrazed riparian habitat in Idaho. USDA For. Serv. Res. Pap. INT-425.

Medin, D.E., and W.P. Clary. 1991. Breeding bird populations in a grazed and ungrazed riparian habitat in Nevada. USDA For. Serv. Res. Pap. INT-441.

Medvin, M.M., M.D. Beecher, and S.J. Andelman. 1987. Extra adults at the nest in Barn Swallows. Condor 89:179-182.

Melillo, J.M., C.A. Palm, R.A. Houghton, G.M. Woodwell, and N. Myers. 1986. A comparison of two recent estimates of disturbance in tropical forests. Environ. Conserv. 12:37-40.

Metz, K.J. 1991. The enigma of multiple nest building by male Marsh Wrens. Auk 108:170-173.

Mewaldt, L.R., and J.R. King. 1985. Breeding site faithfulness, reproductive biology, and adult survivorship in an isolated population of Cassin's Finches. Condor 87:494-510.

Middleton, A.L.A. 1991. Failure of Brown-headed Cowbird parasitism in nests of the American Goldfinch. J. Field Ornithol. 62:200-203.

Milchunas, D.G., O.E. Sala, and W.K. Lauenroth. 1988. A generalized model of the effects of grazing by large herbivores on grassland community structure. Am. Nat. 132:87-106.

Miller, E., and D.R. Miller. 1980. Snag use by birds. In: R.M. DeGraff and N.G. Tilghman (eds.), Workshop Proceedings: Management of Western Forests and Grasslands for Nongame Birds. USDA For. Serv. Gen. Tech. Rep. INT-86, pp. 337-356.

Miller, J.H., and M.T. Green. 1987. Distribution, status, and origin of Water Pipits breeding in California. Condor 89:788-797.

Miller, R.S., and R.W. Nero. 1983. Hummingbird-sapsucker associations in northern climates. Can. J. Zool. 61:1540-1546.

Miller, S.J., and D.W. Inouye. 1983. Roles of the wing whistle in the territorial behaviour of male Broad-tailed Hummingbirds *(Selasphorus platycercus)*. Anim. Behav. 31:689-700.

Mock, P.J. 1991. Daily allocation of time and energy of Western Bluebirds feeding nestlings. Condor 93:598-611.

Mock, P.J., M. Khubesrian, and D.M. Larcheveque. 1991. Energetics of growth and maturation in sympatric passerines that fledge at different ages. Auk 108:34-41.

Moore, K.R., and C.J. Henny. 1983. *Accipiter* nest sites in Oregon. Raptor Res. 17:65-76.

Moore, R.L. 1992. Breeding birds in old-growth forests, snag management for birds. Unpublished Final Report, Forest Plan Monitoring Project, Gallatin Natl. For., Bozeman, Montana.

Moore, W.S., and R.A. Dolbeer. 1989. The use of banding recovery data to estimate dispersal rates and gene flow in avian species: case studies in the Red-winged Blackbird and Common Grackle. Condor 91:242-253.

Moriarty, L.J. 1965. A study of the breeding biology of the Chestnut-collared Longspur *(Calcarius ornatus)* in northeastern South Dakota. South Dakota Bird Notes 17:76-79.

Morimoto, D.C., and F.E. Wasserman. 1991. Intersexual and interspecific differences in the foraging behavior of Rufous-sided Towhees, Common Yellowthroats and Prairie Warblers in the pine barrens of southeastern Massachusetts. J. Field Ornithol. 62:436-449.

Morrison, M.L. 1980. Seasonal aspects of the predatory behavior of Loggerhead Shrikes. Condor 82:296-300.

Morrison, M.L. 1981. The structure of western warbler assemblages: analysis of foraging and habitat selection in Oregon. Auk 98:578-588.

Morrison, M.L. 1983. Analysis of geographic variation in the Townsend's Warbler. Condor 85:385-391.

Morrison, M.L., I.C. Timossi, and K.A. With. 1987. Development and testing of linear regression models predicting bird-habitat relationships. J. Wildl. Manage. 51:247-253.

Morse, D.H. 1989. American Warblers: An Ecological and Behavioral Perspective. Harvard Univ. Press, Cambridge, MA.

Morton, E.S. 1992. What do we know about the future of migrant landbirds? In: J.M. Hagan, III and D.W. Johnston (eds.), Ecology and Conservation of Neotropical Migrant Landbirds. Smithsonian Inst. Press, Washington, DC, pp. 579-589.

Morton, E.S., and K.C. Derrickson. 1990. The biological significance of age-specific return schedules in breeding Purple Martins. Condor 92:1040-1050.

Morton, E.S., L. Forman, and M. Braun. 1990. Extrapair fertilizations and the evolution of colonial breeding in Purple Martins. Auk 107:275-283.

Morton, M.L. 1991. Postfledging dispersal of Green-tailed Towhees to a subalpine meadow. Condor 93:466-468.

Morton, M.L. 1992. Effects of sex and birth date on premigration biology, migration schedules, return rates and natal dispersal in the Mountain White-crowned Sparrow. Condor 94:117-133.

Morton, M.L., and M.E. Pereyra. 1985. The regulation of egg temperatures and attentiveness patterns in the Dusky Flycatcher *(Empidonax oberholseri)*. Auk 102:25-37.

Mosconi, S.L., and R.L. Hutto. 1982. The effect of grazing on the land birds of a western Montana riparian habitat. In: J.M. Peek and P.D. Dalke (eds.), Wildlife-Livestock Relationships Symp. Forest Wildlife Range Exp. Station, Univ. Idaho, Moscow, pp. 221-233.

Mountjoy, D.J., and R.J. Robertson. 1988a. Why are waxwings "waxy"? Delayed plumage maturation in the Cedar Waxwing. Auk 105:61-69.

Mountjoy, D.J., and R.J. Robertson. 1988b. Nest-construction tactics in the Cedar Waxwing. Wilson Bull. 100:128-130.

Mueller, H.C., D.D. Berger, and G. Allez. 1977. The periodic invasions of Goshawks. Auk 94:652-663.

Mulvihill, R.S., R.C. Leberman, and D.S. Wood. 1992. A possible relationship between reversed sexual size dimorphism and reduced male survivorship in the Ruby-throated Hummingbird. Condor 94:480-489.

Munson, E.S. 1992. Influence of nest cover on habitat selection in Clay-colored Sparrows. Wilson Bull. 104:525-529.

Murphy, M.T. 1988. Comparative reproductive biology of kingbirds *(Tyrannus* spp.) in eastern Kansas. Wilson Bull. 100:357-376.

Myers, N. 1980. Conversion of Tropical Moist Forests. National Acad. Sci., Washington, DC.

National Fish and Wildlife Foundation. 1991. Neotropical Migratory Bird Conservation Program, Partners in Flight. Washington, DC.

National Geographic Society. 1990. Birds of North America, second edition. National Geographic Soc., Washington, DC.

Nelson, H.K., and H.F. Duebbert. 1974. New concepts regarding the production of waterfowl and other gamebirds in areas of diversified agriculture. Proc. Intl. Congr. Game Biol. 11:385-394.

Neudorf, D.L., and S.G. Sealy. 1992. Reactions of four passerine species to threats of predation and cowbird parasitism: enemy recognition or generalized responses? Behaviour 123:84-105.

Newman, G.A. 1970. Cowbird parasitism and nesting success of Lark Sparrows in southern Oklahoma. Wilson Bull. 82:304-309.

Nolan, V., Jr., and E.D. Ketterson. 1990. Timing of autumn migration and its relation to winter distribution in Dark-eyed Juncos. Ecology 71:1267-1278.

Noon, B.R. 1981. The distribution of an avian guild along a temperate elevational gradient: the importance and expression of competition. Ecol. Monogr. 51:105-124.

Ohlendorf, H.M. 1974. Competitive relationships among kingbirds *(Tyrannus)* in Trans-Pecos Texas. Wilson Bull. 86:357-373.

Olson, J.M. 1991. Thermal relations of nestling Red-winged Blackbirds in southeastern Michigan. Auk 108:711-716.

Olson, S.L. 1984. Density and distribution, nest site selection, and activity of the Mountain Plover on the Charles M. Russell National Wildlife Refuge. M.S. Thesis, Univ. of Montana, Missoula.

Olson, T.E., and M.V. Gray. 1989. Characteristics of the Least Bell's Vireo nest sites along the Santa Ynez River, Santa Barbara County. In: Proceedings of the California Riparian Systems Conference. USDA Forest Serv. Gen. Tech. Rep. PSW-110, pp. 278-284.

Olson, T.E., and F.L. Knopf. 1986. Naturalization of Russian-olive in the western United States. West. J. Appl. For. 1:65-69.

Olson-Edge, S.L., and W.D. Edge. 1987. Density and distribution of the Mountain Plover on the Charles M. Russell National Wildlife Refuge. Prairie Nat. 19:233-238.

Orians, G. H. 1980. Some Adaptations of Marsh-nesting Blackbirds. Monogr. Pop. Biol. No. 14, Princeton Univ. Press, NJ.

Orians, G.H. 1985. Blackbirds of the Americas. Univ. Washington Press, Seattle.

Oring, L.W., and D.B. Lank. 1986. Environment and experience: Impact on a polyandrous population of Spotted Sandpipers. In: D.I. Rubenstein and R.W. Wrangham (eds.), Ecological Aspects of Social Evolution. Princeton Univ. Press, NJ, pp. 21-42.

Ortega, C.P., and A. Cruz. 1991. A comparative study of cowbird parasitism in Yellow-headed Blackbirds and Red-winged Blackbirds. Auk 108:16-24.

Owens, R.A., and M.T. Myres. 1973. Effects of agriculture upon populations of native passerine birds of an Alberta fescue grassland. Can. J. Zool. 51:697-713.

Paige, L.C. 1990. Population Trends of Songbirds in Western North America. Thesis, Univ. Montana, Missoula.

Paton, D.C., and F.L. Carpenter. 1984. Peripheral foraging by territorial Rufous Hummingbirds: defense by exploitation. Ecology 65:1808-1819.

Patterson, C.B. 1991. Relative parental investment in the Red-winged Blackbird. J. Field Ornithol. 62:1-18.

Payne, R.B., and L.L. Payne. 1990. Survival estimates of Indigo Buntings: comparison of banding recoveries and local observations. Condor 92:938-946.

Perritt, J.E., and L.B. Best. 1989. Effects of weather on the breeding ecology of Vesper Sparrows in Iowa crop fields. Am. Mid. Nat. 121:355-360.

Petersen, K.L., and L.B. Best. 1987. Effects of prescribed burning on nongame birds in a sagebrush community. Wildl. Soc. Bull. 15:317-329.

Petersen, K.L., and L.B. Best. 1991. Nest-site selection by Sage Thrashers in southeastern Idaho. Great Basin Natur. 51:261-266.

Peterson, R.T. 1990. A Field Guide to Western Birds, third edition. Houghton Mifflin Co., Boston.

Peterson, S.R. 1982. A preliminary survey of forest bird communities in northern Idaho. Northwest Sci. 56:287-298.

Petrinovich, L., and T.L. Patterson. 1983. The White-crowned Sparrow: reproductive success (1975–1980). Auk 100:811-825.

Pickett, P.E., S.J. Maxson, and L.W. Oring. 1988. Interspecific interactions of Spotted Sandpipers. Wilson Bull. 100:297-302.

Picman, J. 1984. Experimental study on the role of intra- and inter-specific competition in the evolution of nest-destroying behavior in Marsh Wrens. Can. J. Zool. 62:2353-2356.

Picman, J., and A.K. Picman. 1980. Destruction of nests by the Short-billed Marsh Wren. Condor 82:176-179.

Pinkowski, B.C. 1979. Annual productivity and its measurement in a multi-brooded passerine, the Eastern Bluebird. Auk 96:562-572.

Place, A.R., and E.W. Stiles. 1992. Living off the wax of the land: bayberries and Yellow-rumped Warblers. Auk 109:334-345.

Platt, J.B. 1976. Sharp-shinned Hawk nesting and nest site selection in Utah. Condor 78:102-103.

Pleasants, B.Y. 1979. Adaptive significance of the variable dispersion pattern of breeding Northern Orioles. Condor 81:28-34.

Pleszczynska, W.K. 1978. Microgeographic prediction of polygyny in Lark Bunting. Science 201:935-937.

Pleszczynska, W.K., and R. Hansell. 1980. Polygyny and decision theory: testing of a model in Lark Buntings (*Calamospiza melanocorys*). Am. Nat. 116:821-830.

Pletschet, S.M. 1987. Habitat preferences and interspecific competition: Red-eyed and Warbling Vireos. M.S. Thesis, Univ. of Montana, Missoula.

Plissner, J.H., and P.A. Gowaty. 1988. Evidence of reproductive error in adoption of nestling Eastern Bluebirds *(Sialia sialis)*. Auk 105:575-578.

Poole, A. 1989. Ospreys: A Natural and Unnatural History. Cambridge Univ. Press, Cambridge.

Porter, D.K., M.S. Strong, J.B. Giezentanner, and R.A. Ryder. 1975. Nest ecology, productivity and growth of the Loggerhead Shrike on the short-grass prairie. Southwest. Natur. 19:429-436.

Powell, A.N., and F.J. Cuthbert. 1993. Augmenting small populations of plovers: An assessment of cross-fostering and captive-rearing. Cons. Biol. 7:160-168.

Power, H.W. 1980. The Foraging Behavior of Mountain Bluebirds with Emphasis on Sexual Foraging Differences. Ornithol. Monogr. No. 28.

Prior, K.A. 1990. Turkey Vulture food habits in southern Ontario. Wilson Bull. 102:706-710.

Procter-Gray, E. 1991. Female-like plumage of subadult male American Redstarts does not reduce aggression from other males. Auk 108:872-879.

Quay, W.B. 1989. Insemination of Tennessee Warblers during spring migration. Condor 91:660-670.

Quinn, M.A., and D.D. Walgenbach. 1990. Influence of grazing history on the community structure of grasshoppers of a mixed-grass prairie. Environ. Entomol. 19:1756-1766.

Raley, C.M., and S.H. Anderson. 1990. Availability and use of arthropod food resources by Wilson's Warblers and Lincoln's Sparrows in southeastern Wyoming. Condor 92:141-150.

Ramsden, D.J., L.J. Lyon, and G.L. Halvorson. 1979. Small bird populations and feeding habitats—western Montana in July. Am. Birds 33:

Rappole, J.H., E.S. Morton, T.E. Lovejoy, III, and J.L. Ruos. 1983. Nearctic Avian Migrants in the Neotropics. USDI Fish Wildl. Serv., Washington, DC.

Ratcliffe, D. 1980. The Peregrine Falcon. Buteo Books, Vermillion, SD.

Redmond, R.L., and D.A. Jenni. 1982. Natal philopatry and breeding area fidelity of Long-billed Curlews *(Numenius americanus)*: patterns and evolutionary consequences. Behav. Ecol. Sociobiol. 10:277-279.

Redmond, R.L., and D.A. Jenni. 1986. Population ecology of the Long-billed Curlew *(Numenius americanus)* in western Idaho. Auk 103:755-767.

Rendell, W.B., and R.J. Robertson. 1989. Nest-site characteristics, reproductive success and cavity availability for Tree Swallows breeding in natural cavities. Condor 91:875-885.

Rendell, W.B., and R.J. Robertson. 1990. Influence of forest edge on nest-site selection by Tree Swallows. Wilson Bull. 102:634-644.

Renken, R.B., and J.J. Dinsmore. 1987. Nongame bird communities on managed grasslands in North Dakota. Can. Field-Nat. 101:551-557.

Restani, M. 1991. Resource partitioning among three *Buteo* species in the Centennial Valley, Montana. Condor 93:1007-1010.

Reynolds, R.T. 1983. Management of western coniferous forest habitat for nesting Accipiter hawks. USDA Forest Serv. Gen Tech. Rep. RM-102.

Reynolds, R.T., and B.D. Linkhart. 1987a. Fidelity to territory and mate in Flammulated Owls. In: R.W. Nero, R.J. Clark, R.J. Knapton, and R.H. Hamre (eds.), Biology and Conservation of Northern Forest Owls. USDA Forest Serv. Gen. Tech. Rep. RM-142, pp. 234-238.

Reynolds, R.T., and B.D. Linkhart. 1987b. The nesting biology of Flammulated Owls in Colorado. In: R.W. Nero, R.J. Clark, R.J. Knapton, and R.H. Hamre (eds.), Biology and Conservation of Northern Forest Owls. USDA Forest Serv. Gen. Tech. Rep. RM-142, pp. 239-248.

Reynolds, R.T., and E.C. Meslow. 1984. Partitioning of food and niche characteristics of coexisting *Accipiter* during breeding. Auk 101:761-779.

Reynolds, R.T., E.C. Meslow, and H.M. Wight. 1982. Nesting habitat of coexisting *Accipiter* in Oregon. J. Wildl. Manage. 46:124-138.

Reynolds, T.D. 1981. Nesting of the Sage Thrasher, Sage Sparrow, and Brewer's Sparrow in southeastern Idaho. Condor 83:61-64.

Reynolds, T.D., and T.D. Rich. 1978. Reproductive ecology of the Sage Thrasher *(Oreoscoptes montanus)* on the Snake River Plain in southeastern Idaho. Auk 95:580-582.

Reynolds, T.D., and C.H. Trost. 1981. Grazing, crested wheatgrass, and bird populations in southeastern Idaho. Northwest Sci. 55:225-234.

Rice, W.R. 1982. Acoustical location of prey by the Marsh Hawk: adaptation to concealed prey. Auk 99:403-413.

Rich, A.C., D.S. Dobkin, and L.J. Niles. Defining forest fragmentation by corridor width: the influence of narrow forest-dividing corridors on forest-nesting Neotropical migrant landbirds in southern New Jersey. Cons. Biol., in press.

Rich, T.D. 1986. Habitat and nest site selection by Burrowing Owls in the sagebrush steppe of Idaho. J. Wildl. Manage, 50:548-555.

Rich, T.D., and S.I. Rothstein. 1985. Sage Thrashers reject cowbird eggs. Condor 87:561-562.

Ricklefs, R.E. 1972. Latitudinal variation in breeding productivity of the Rough-winged Swallow. Auk 89:826-836.

Rising, J.D. 1988. Geographic variation in sex ratios and body size in wintering flocks of Savannah Sparrows *(Passerculus sandwichensis)*. Wilson Bull. 100:183-203.

Ritchison, G. 1988. Responses of Yellow-breasted Chats to the songs of neighboring and non-neighboring conspecifics. J. Field Ornithol. 59:37-42.

Ritchison, G. 1991. The flight songs of Common Yellowthroats: description and causation. Condor 93:12-18.

Robbins, C.S. 1980. Effects of forest fragmentation on breeding bird populations in the Piedmont of the mid-Atlantic region. Atl. Natur. 33:31-36.

Robbins, C.S., D. Bystrak, and P.H. Geissler. 1986. The breeding bird survey: Its first 15 years, 1965–1979. USDI Fish Wildl. Serv. Res. Publ. No. 157.

Robbins, C.S., D.K. Dawson, and B.A. Dowell. 1989. Habitat Area Requirements of Breeding Forest Birds of the Middle Atlantic States. Wildl. Monogr. No. 103.

Robbins, C.S., J.R. Sauer, R.S. Greenberg, and S. Droege. 1989. Population declines in North American birds that migrate to the neotropics. Proc. Natl. Acad. Sci. USA 86:7658-7662.

Robinson, S.K. 1981. Ecological relations and social interactions of Philadelphia and Red-eyed Vireos. Condor 83:16-26.

Robinson, S.K. 1992. Population dynamics of breeding Neotropical migrants in a fragmented Illinois landscape. In: J.M. Hagan, III, and D.W. Johnston (eds.), Ecology and Conservation of Neotropical Migrant Landbirds. Smithsonian Inst. Press, Washington, DC, pp. 408-418.

Rohwer, S., and J. Manning. 1990. Differences in timing and number of molts for Baltimore and Bullock's Orioles: implications to hybrid fitness and theories of delayed plumage maturation. Condor 92:125-140.

Rohwer, S., C.D. Spaw, and E. Roskaft. 1989. Costs to Northern Orioles of puncture-ejecting parasitic cowbird eggs from their nests. Auk 106:734-738 Rosenberg, K.V., and M.G. Raphael. 1986. Effects of forest fragmentation on vertebrates in Douglas-fir forests. In: J. Verner, M.L. Morrison, and C.J. Ralph (eds.), Wildlife 2000: Modeling Habitat Relationships of Terrestrial Vertebrates. Univ. Wisconsin Press, Madison, pp. 263-272.

Rosenfield, R.N. 1984. Nesting biology of Broad-winged Hawks in Wisconsin. Raptor Res. 18:6-9.

Rosenfield, R.N., J. Bielefeldt, and J. Cary. 1991. Copulatory and other pre-incubation behaviors of Cooper's Hawks. Wilson Bull. 103:656-660.

Rotenberry, J.T., and J.A. Wiens. 1989. Reproductive biology of shrubsteppe passerine birds: geographical and temporal variation in clutch size, brood size, and fledging success. Condor 91:1-14, 1989.

Rothfels, M., and M.R. Lein. 1983. Ecological interaction of sympatric Red-tailed and Swainson's Hawks. Can. J. Zool. 61:60-64.

Rothstein, S.I., 1976a. Cowbird parasitism of the Cedar Waxwing and its evolutionary implications. Auk 93:498-509.

Rothstein, S.I. 1976b. Experiments on defenses Cedar Waxwings use against cowbird parasitism. Auk 93:675-691.

Ryan, M.R., and R.B. Renken. 1987. Habitat use by breeding Willets in the northern Great Plains. Wilson Bull. 99:175-189.

Ryan, M.R., R.B. Renken, and J.J. Dinsmore. 1984. Marbled Godwit habitat selection in the northern prairie region. J. Wildl. Manage. 48:1206-1218.

Ryder, R.A. 1980. Effects of grazing on bird habitats. In: R.M. DeGraff and N.G. Tilghman (eds.), Workshop Proceedings: Management of Western Forests and Grasslands for Nongame Birds. USDA For. Serv. Gen. Tech. Rep. INT-86, pp. 51-66.

Saab, V.A., and C.R. Groves. 1992. Idaho's migratory landbirds: description, habitats, and conservation. Nongame Wildl. Leaflet No. 10. Idaho Wildl., Summer 1992:11-26.

Sabo, S.R., and R.T. Holmes. 1983. Foraging niches and the structure of forest bird communities in contrasting montane habitats. Condor 85:121-138.

Sader, S.A., and A.T. Joyce. 1988. Deforestation rates and trends in Costa Rica, 1940 to 1983. Biotropica 20:11-19.

Sakai, H.F., and B.R. Noon. 1991. Nest-site characteristics of Hammond's and Pacific-slope Flycatchers in northwestern California. Condor 93:563-574.

Salomonson, M.G., and R.P. Balda. 1977. Winter territoriality of Townsend's Solitaires *(Myadestes townsendi)* in a pinon-juniper-ponderosa pine ecotone. Condor 79:148-161.

Salt, G.W. 1957. An analysis of avifaunas in the Teton Mountains and Jackson Hole, Wyoming. Condor 59:373-393.

Samson, F.B. 1976. Territory, breeding density, and fall departure in Cassin's Finch. Auk 93:477-497.

Sanders, S.D., and M.A. Flett. 1989. Montane riparian habitat and Willow Flycatchers: threats to a sensitive environment and species. In: Proceedings of the California Riparian Systems Conference. USDA Forest Serv. Gen. Tech. Rep. PSW-110, pp. 262-266.

Sanderson, H.R, E.L. Bull, and P.J. Edgerton. 1980. Bird communities in mixed conifer forests of the interior Northwest. In: R.M. DeGraff and N.G. Tilghman (eds.), Workshop Proceedings: Management of Western Forests and Grasslands for Nongame Birds. USDA For. Serv. Gen. Tech. Rep. INT-86, pp. 224-237.

Sauer, J.R., and S. Droege. 1990. Recent population trends of the Eastern Bluebird. Wilson Bull. 102:239-252.

Sauer, J.R., and S. Droege. 1992. Geographic patterns in population trends of Neotropical migrants in North America. In: J.M. Hagan, III, and D.W. Johnston (eds.), Ecology and Conservation of Neotropical Migrant Landbirds. Smithsonian Inst. Press, Washington, DC, pp. 26-42.

Schaeff, C., and J. Picman. 1988. Destruction of eggs by Western Meadowlarks. Condor 90:935-937.

Schmutz, J.K. 1989. Hawk occupancy of disturbed grasslands in relation to models of habitat selection. Condor 91:362-371.

Schmutz, J.K., R.W. Fyfe, U. Banasch, and H. Armbruster. 1991. Routes and timing of migration of falcons banded in Canada. Wilson Bull. 103:44-58.

Schroeder, M.H., and D.L. Sturges. 1975. The effect on the Brewer's Sparrow of spraying big sagebrush. J. Range Manage. 28:294-297.

Schroeder, R.L. 1982a. Habitat suitability index models: Yellow Warbler. U.S. Dept. Int., Fish Wildl. Serv. FWS/OBS-82/10.27.

Schroeder, R.L. 1982b. Habitat suitability index models: Yellow-headed Blackbird. U.S. Dept. Int., Fish Wildl. Serv. FWS/OBS-82/10.26.

Schulz, T.T., and W.C. Leininger. 1991. Nongame wildlife communities in grazed and ungrazed montane riparian sites. Great Basin Natur. 51:286-292. Scott, V.E., and G.L. Crouch. 1988a. Summer birds and mammals of aspen-conifer forests in west-central Colorado. USDA For. Serv. Res. Pap. RM-280.

Scott, V.E., and G.L. Crouch. 1988b. Breeding birds in uncut aspen and 6- to 10-year-old clearcuts in southwestern Colorado. USDA For. Serv. Res. Note RM-485.

Sealy, S.G. 1974. Ecological segregation of Swainson's and Hermit Thrushes on Langara Island, British Columbia. Condor 76:350-351.

Sealy, S.G. 1978. Possible influence of food on egg-laying and clutch size in the Black-billed Cuckoo. Condor 80:103-104.

Sealy, S.G. 1980a. Breeding biology of Orchard Orioles in a new population in Manitoba. Can. Field-Natur. 94:154-158.

Sealy, S.G. 1980b. Reproductive responses of Northern Orioles to a changing food supply. Can. J. Zool. 58:221-227.

Sealy, S.G. 1985. Erect posture of the young Black-billed Cuckoo: an adaptation for early mobility in a nomadic species. Auk 102:889-892.

Sealy, S.G. 1992. Removal of Yellow Warbler eggs in association with cowbird parasitism. Condor 94:40-54.

Sealy, S.G., K.A. Hobson, and J.V. Briskie. 1989. Responses of Yellow Warblers to experimental intraspecific brood parasitism. J. Field Ornithol. 60:224-229.

Secunda, R.C., and T.W. Sherry. 1991. Polyterritorial polygyny in the American Redstart. Wilson Bull. 103:190-203.

Sedgwick, J.A. 1993. Reproductive ecology of Dusky Flycatchers in western Montana. Wilson Bull. 105:84-92.

Sedgwick, J.A., and F.L. Knopf. 1987. Breeding bird response to cattle grazing of a cottonwood bottomland. J. Wildl. Manage. 51:230-237.

Sedgwick, J.A., and F.L. Knopf. 1989. Regionwide polygyny in Willow Flycatchers. Condor 91:473-475.

Sedgwick, J.A., and F.L. Knopf. 1990. Habitat relationships and nest site characteristics of cavity-nesting birds in cottonwood floodplains. J. Wildl. Manage. 54:112-124.

Sherrod, S.K. 1978. Diets of North American Falconiformes. Raptor Res. 12:49-121.

Sherry, T.W., and R.T. Holmes. 1988. Habitat selection by breeding American Redstarts in response to a dominant competitor, the Least Flycatcher. Auk 105:350-364.

Sherry, T.W., and R.T. Holmes. 1992. Population fluctuations in a long-distance Neotropical migrant: Demographic evidence for the importance of breeding season events in the American Redstart. In: J.M. Hagan, III and D.W. Johnston (eds.), Ecology and Conservation of Neotropical Migrant Landbirds. Smithsonian Inst. Press, Washington, DC, pp. 431-442.

Shields, W.M. 1984. Factors affecting nest and site fidelity in Adirondack Barn Swallows *(Hirundo rustica)*. Auk 101:780-789.

Shields, W.M. 1990. Information centers and coloniality in Cliff Swallows: statistical design and analysis—a comment. Ecology 71:401-405.

Short, H.L. 1984. Habitat suitability index models: Brewer's Sparrow. U.S. Dept. Int., Fish Wildl. Serv. FWS/OBS-82/10.83.

Shy, E. 1984. Habitat shift and geographical variation in North American tanagers (Thraupinae: *Piranga*). Oecologia 63:281-285.

Sieg, C.H., and D.M. Becker. 1990. Nest-site habitat selected by Merlins in southeastern Montana. Condor 92:688-694.

Simmons, R. 1988. Honest advertising, sexual selection, courtship displays, and body condition of polygynous male harriers. Auk 105:303-307.

Sims, P.L. 1988. Grasslands. In: M.G. Barbour and W.D. Billings (eds.), North American Terrestrial Vegetation. Cambridge Univ. Press, NY, pp. 265-286.

Skagen, S.K. 1987. Hatching asynchrony in American Goldfinches: an experimental study. Ecology 68:1747-1759.

Skinner, N.G. 1989. Seasonal avifauna use of burned and unburned lodgepole pine forest ecotones. Thesis, Univ. Montana, Missoula.

Smith, D.G., J.R. Murphy, and N.D. Woffindin. 1981. Relationships between jackrabbit abundance and Ferruginous Hawk reproduction. Condor 83:52-56.

Smith, H.G., and R. Montgomerie. 1991. Nestling American Robins compete with siblings by begging. Behav. Ecol. Sociobiol. 29:307-312.

Smith, J.N.M., and J.R. Merkt. 1980. Development and stability of single-parent family units in the Song Sparrow. Can. J. Zool. 58:1869-1875.

Smith, J.N.M., Y. Yom-Tov, and R. Moses. 1982. Polygyny, male parental care, and sex ratio in Song Sparrows: an experimental study. Auk 99:555-564.

Smith, K.G. 1980. Nongame birds of the Rocky Mountain spruce-fir forests and their management. In: R.M. DeGraff and N.G. Tilghman (eds.), Workshop Proceedings: Management of Western Forests and Grasslands for Nongame Birds. USDA For. Serv. Gen. Tech. Rep. INT-86, pp. 258-279.

Smith, K.G. 1982. Drought-induced changes in avian community structure along a montane sere. Ecology 63:952-961.

Smith, K.G., and D.C. Andersen. 1982. Food, predation, and reproductive ecology of the Dark-eyed Junco in northern Utah. Auk 99:650-661.

Smith, T.M., and H.H. Shugart. 1987. Territory size variation in the Ovenbird: the role of habitat structure. Ecology 68:695-704.

Smith, W.J., J. Pawlukiewicz, and S.T. Smith. 1978. Kinds of activities correlated with singing patterns of the Yellow-throated Vireo. Anim. Behav. 26:862-884.

Smith, W.W. 1987. The Baird's Sparrow in Alberta. In: Endangered Species in the Prairie Provinces, Proceedings of the Workshop. Prov. Mus. Alberta Nat. Hist. Occ. Pap. No. 9, pp. 277-279.

Snapp, B.D. 1976. Colonial breeding in the Barn Swallow *(Hirundo rustica)* and its adaptive significance. Condor 78:471-480.

Sodhi, N.S., I.G. Warkentin, P.C. James, and L.W. Oliphant. 1991. Effects of radiotagging on breeding Merlins. J. Wildl. Manage. 55:613-616.

Sordahl, T.A. 1979. Vocalizations and behavior of the Willet. Wilson Bull. 91:551-574.

Sousa, P.J. 1982. Habitat suitability index models: Veery. U.S. Dept. Int., Fish Wildl. Serv. FWS/OBS-82/10.22.

Sousa, P.J. 1983. Habitat suitability index models: Lewis' Woodpecker. U.S. Dept. Int., Fish Wildl. Serv. FWS/OBS-82/10.32.

Sousa, P.J., and W.N. McDonal. 1983. Habitat suitability index models: Baird's Sparrow. U.S. Dept. Int., Fish Wildl. Serv. FWS/OBS-82/10.44.

South Dakota Ornithologists' Union. 1991. The Birds of South Dakota, second edition. Northern State Univ. Press, Aberdeen, SD.

Squires, J.R., S.H. Anderson, and R. Oakleaf. 1989. Food habits of nesting Prairie Falcons in Campbell County, Wyoming. J. Raptor Res. 23:157-161.

Squires, J.R., S.H. Anderson, and R. Oakleaf. 1993. Home range size and habitat-use patterns of nesting Prairie Falcons near oil developments in northeastern Wyoming. J. Field Ornithol. 64:1-10.

St. Louis, V.L., and L. Breebaart. 1991. Calcium supplements in the diet of nestling Tree Swallows near acid sensitive lakes. Condor 93:286-294.

Steenhof, K., M.N. Kochert, and J.H. Doremus. 1983. Nesting of subadult Golden Eagles in southwestern Idaho. Auk 100:743-747.

Steidl, R.J., and C.R. Griffin. 1991. Growth and brood reduction of mid-Atlantic Coast Ospreys. Auk 108:363-370.

Steidl, R.J., C.R. Griffin, and L.J. Niles. 1991. Contaminant levels of Osprey eggs and prey reflect regional differences in reproductive success. J. Wildl. Manage. 55:601-608.

Steidl, R.J., C.R. Griffin, L.J. Niles, and K.E. Clark. 1991. Reproductive success and eggshell thinning of a reestablished Peregrine Falcon population. J. Wildl. Manage. 55:294-299.

Stendell, R.C., D.S. Gilmer, N.A. Coon, and D.M. Swineford. 1988. Organochlorine and mercury residues in Swainson's and Ferruginous Hawk eggs collected in North and South Dakota, 1974-79. Environ. Monitor. Assess. 10:37-41.

Stephens, D.A., and S.H. Sturts. 1991. Idaho Bird Distribution. Idaho Mus. Natur. Hist. Spec. Publ. No. 11, Pocatello, ID.

Stewart, R.E. 1975. Breeding Birds of North Dakota. Tri-College Center Environ. Studies, Fargo, ND.

Stewart, R.M., R.P. Henderson, and K. Darling. 1977. Breeding ecology of the Wilson's Warbler in the high Sierra Nevada, California. Living Bird 16:83-102.

Stoddard, P.K., M.D. Beecher, C.L. Horning, and M.S. Willis. 1990. Strong neighbor-stranger discrimination in Song Sparrows. Condor 92:1051-1056.

Stromberg, M.R., and P.B. Johnsen. 1990. Hummingbird sweetness preferences: taste or viscosity? Condor 92:606-612.

Stutchbury, B.J. 1988. Evidence that Bank Swallow colonies do not function as information centers. Condor 90:953-955.

Stutchbury, B.J. 1991. Coloniality and breeding biology of Purple Martins *(Progne subis hesperia)* in saguaro cacti. Condor 93:666-675.

Swanson, D.L. 1991. Seasonal adjustments in metabolism and insulation in the Dark-eyed Junco. Condor 93:538-545.

Szaro, R.C. 1991. Wildlife communities of southwestern riparian ecosystems. In: J.E. Rodiek and E.G. Bolen (eds.), Wildlife and Habitats in Managed Landscapes. Island Press, Washington, DC, pp. 174-201.

Szaro, R.C., and R.P. Balda. 1979. Bird community dynamics in a ponderosa pine forest. Stud. Avian Biol. No. 3.

Szaro, R.C., and R.P. Balda. 1986. Relationships among weather, habitat structure, and ponderosa pine forest birds. J. Wildl. Manage. 50:253-260.

Szaro, R.C., J.D. Brawn, and R.P. Balda. 1990. Yearly variation in resource-use behavior by ponderosa pine forest birds. Stud. Avian Biol. 13:226-236.

Tamm, S. 1989. Display behavior of male Calliope Hummingbirds during the breeding season. Condor 91:272-279.

Tate, J., Jr. 1973. Methods and annual sequence of foraging by the sapsucker. Auk 90:840-856.

Taylor, D.M., and C.D. Littlefield. 1986. Willow Flycatcher and Yellow Warbler response to cattle grazing. Am. Birds 40:1169-1173.

Temeles, E.J. 1989. Effect of prey consumption on foraging activity of Northern Harriers. Auk 106:353-357.

Temple, S.A. (ed.). 1978. Endangered Birds: Management Techniques for Preserving Threatened Species. Univ. Wisconsin Press, Madison.

Temple, S.A., and J.R. Cary. 1988. Modeling dynamics of habitat-interior bird populations in fragmented landscapes. Conserv. Biol. 2:340-347.

Terborgh, J.W. 1980. The conservation status of Neotropical migrants: Present and future. In: A. Keast and E.S. Morton (eds.), Migrant Birds in the Neotropics: Ecology, Behavior, Distribution, and Conservation. Smithsonian Inst. Press, Washington, DC, pp. 21-30.

Terborgh, J. 1989. Where Have All the Birds Gone? Princeton Univ. Press, NJ.

Thomas, J.W., C. Maser, and J.E. Rodiek. 1979. Wildlife habitats in managed rangelands—the Great Basin of southeastern Oregon. Riparian Zones. USDA For. Serv. Gen. Tech. Rep. PNW-80.

Thompson, C.F. 1977. Experimental removal and replacement of territorial male Yellow-breasted Chats. Auk 94:107-113.

Thompson, C.F., and V. Nolan, Jr. 1973. Population biology of the Yellow-breasted Chat (*Icteria virens* L.) in southern Indiana. Ecol. Monogr. 43:145-171.

Thompson, L.S. 1978. Species abundance and habitat relations of an insular montane avifauna. Condor 80:1-14.

Thompson, W.L., R.H. Yahner, and G.L. Storm. 1990. Winter use and habitat characteristics of vulture communal roosts. J. Wildl. Manage. 54:77-83.

Threlfall, W., and J.R. Blacquiere. 1982. Breeding biology of the Fox Sparrow in Newfoundland. J. Field Ornithol. 53:235-239.

Tobalske, B.W. 1992. Evaluating habitat suitability using relative abundance and fledging success of Red-naped Sapsuckers. Condor 94:550-553.

Tobalske, B.W., R.C. Shearer, and R.L. Hutto. 1991. Bird populations in logged and unlogged western larch/Douglas-fir forest in northwestern Montana. USDA For. Serv. Res. Pap. INT-442.

Toland, B.R. 1987. The effect of vegetative cover on foraging strategies, hunting success and nesting distribution of American Kestrels in central Missouri. J. Raptor Res. 21:14-20.

Torgersen, T.R., R.R. Mason, and R.W. Campbell. 1990. Predation by birds and ants on two forest insect pests in the Pacific Northwest. Stud. Avian Biol. 13:14-19.

Tubbs, A.A. 1980. Riparian bird communities of the Great Plains. In: R.M. DeGraff and N.G. Tilghman (eds.), Workshop Proceedings: Management of Western Forests and Grasslands for Nongame Birds. USDA For. Serv. Gen. Tech. Rep. INT-86, pp. 419-433.

Twedt, D.J., W.J. Bleier, and G.M. Linz. 1991. Geographic and temporal variation in the diet of Yellow-headed Blackbirds. Condor 93:975-986.

Tyrrell, E.Q., and R.A. Tyrrell. 1984. Hummingbirds: Their Life and Behavior. Crown, New York.

Van Daele, L.J., and H.A. Van Daele. 1982. Factors affecting the productivity of Ospreys nesting in west-central Idaho. Condor 84:292-299.

Varland, D.E., E.E. Klaas, and T.M. Loughin. 1991. Development of foraging behavior in the American Kestrel. J. Raptor Res. 25:9-17.

Verbeek, N.A.M. 1975a. Northern wintering of flycatchers and residency of Black Phoebes in California. Auk 92:737-749.

Verbeek, N.A.M. 1975b. Comparative feeding behavior of three coexisting tyrannid flycatchers. Wilson Bull. 87:231-240.

Verner, J., and T.A. Larson. 1989. Richness of breeding bird species in mixed-conifer forests of the Sierra Nevada, California. Auk 106:447-463.

Villard, M.-A., K. Freemark, and G. Merriam. 1992. Metapopulation theory and Neotropical migrant birds in temperate forests: An empirical investigation. In: J.M. Hagan, III and D.W. Johnston (eds.), Ecology and Conservation of Neotropical Migrant Landbirds. Smithsonian Inst. Press, Washington, DC, pp. 474-482.

Walsberg, G.E., and C.A. Schmidt. 1992. Effects of variable humidity on embryonic development and hatching success of Mourning Doves. Auk 109:309-314.

Warkentin, I.G., P.C. James, and L.W. Oliphant. 1990. Body morphometrics, age structure, and partial migration of urban Merlins. Auk 107:25-34.

Waser, N.M. 1976. Food supply and nest timing of Broad-tailed Hummingbirds in the Rocky Mountains. Condor 78:133-135.

Watts, B.D. 1991. Effects of predation risk on distribution within and between habitats in Savannah Sparrows. Ecology 72:1515-1519.

Weatherhead, P.J. 1989. Sex ratios, host-specific reproductive success, and impact of Brown-headed Cowbirds. Auk 106:358-366.

Weatherhead, P.J., and S.B. McRae. 1990. Brood care in American Robins: implications for mixed reproductive strategies by females. Anim. Behav. 39:1179-1188.

Weeks, H.P., Jr. 1978. Clutch size variation in the Eastern Phoebe in southern Indiana. Auk 95:656-666.

Wershler, C.R. 1987. The Mountain Plover in Canada. In: Endangered Species in the Prairie Provinces, Proceedings of the Workshop. Prov. Mus. Alberta, Nat. Hist. Occ. Pap. No. 9, pp. 259-261.

West, N.E. 1988. Intermountain deserts, shrub steppes, and woodlands. In: M.G. Barbour and W.D. Billings (eds.), North American Terrestrial Vegetation. Cambridge Univ. Press, NY, pp. 209-230.

Westneat, D.F. 1988. Male parental care and extrapair copulations in the Indigo Bunting. Auk 105:149-160.

Westneat, D.F. 1989. Intensity of nest defense in Indigo Buntings increases with stage and not number of visits. Auk 106:747-749.

Wheeler, A.H. 1992. Reproductive parameters for free ranging American Kestrels (*Falco sparverius*) using nest boxes in Montana and Wyoming. J. Raptor Res. 26:6-9.

Wheelwright, N.T. 1986. The diet of American Robins: an analysis of U.S. Biological Survey records. Auk 103:710-725.

Wheelwright, N.T., and F.B. Dorsey. 1991. Short-term and long-term consequences of predator avoidance by Tree Swallows *(Tachycineta bicolor)*. Auk 108:719-723.

Whicker, A.D., and J.K. Detling. 1988. Ecological consequences of prairie dog disturbance. BioScience 38:778-785.

Whitcomb, R.F., C.S. Robbins, J.F. Lynch, B.L. Whitcomb, M.K. Klimkiewicz, and D. Bystrak. 1981. Effects of forest fragmentation on avifauna of the eastern deciduous forest. In: R.L. Burgess and D.M. Sharpe (eds.), Forest Island Dynamics in Man-dominated Landscapes. Springer-Verlag, NY, pp. 125-205.

Wiedenfeld, D.A. 1992. Foraging in temperate- and tropical-breeding and wintering male Yellow Warblers. In: J.M. Hagan, III and D.W. Johnston (eds.), Ecology and Conservation of Neotropical Migrant Landbirds. Smithsonian Inst. Press, Washington, DC, pp. 321-328.

Wiens, J.A. 1973. Interterritorial habitat variation in Grasshopper and Savannah Sparrows. Ecology 54:877-884.

Wiens, J.A. 1989a. The Ecology of Bird Communities, Vol. 1, Foundations and Patterns. Cambridge Univ. Press, NY.

Wiens, J.A. 1989b. The Ecology of Bird Communities, Vol. 2, Processes and Variations. Cambridge Univ. Press, NY.

Wiens, J.A., and J.T. Rotenberry. 1981. Habitat associations and community structure in shrubsteppe environments. Ecol. Monogr. 51:21-41.

Wiens, J.A., J.T. Rotenberry, and B. Van Horne. 1986. A lesson in the limitations of field experiments: shrubsteppe birds and habitat alteration. Ecology 67:365-376.

Wiens, J.A., B. Van Horne, and J.T. Rotenberry. 1987. Temporal and spatial variations in the behavior of shrubsteppe birds. Oecologia 73:60-70.

Wiens, J.A., B. Van Horne, and J.T. Rotenberry. 1990. Comparisons of the behavior of Sage and Brewer's Sparrows in shrubsteppe habitats. Condor 92:264-266.

Wilbur, S.R., and J.A. Jackson. 1983. Vulture Biology and Management. Univ. of California Press, Berkeley.

Wilcove, D. 1985. Nest predation in forest tracts and the decline of migratory songbirds. Ecology 66:1211-1214.

Wilcove, D. 1990. Empty skies. Nature Conservancy Mag. Jan-Feb 1990:4-13.

Wilcove, D.S., and J.W. Terborgh. 1984. Patterns of population decline in birds. Am. Birds 38:10-13.

Wilcox, L. 1980. Observations of the life history of Willets on Long Island, New York. Wilson Bull. 92:253-258.

Williams, J.B. 1987. Field metabolism and food consumption of Savannah Sparrows during the breeding season. Auk 104:277-289.

Williams, J.B. 1988. Field metabolism of Tree Swallows during the breeding season. Auk 105:706-714.

Williamson, P. 1971. Feeding ecology of the Red-eyed Vireo *(Vireo olivaceus)* and associated foliage-gleaning birds. Ecol. Monogr. 41:129-152.

Willimont, L.A., S.E. Senner, and L. Goodrich. 1988. Fall migration of Ruby-throated Hummingbirds in the northeastern United States. Wilson Bull. 100:482-488.

Wilson, E.O. (ed.). 1988. Biodiversity. Natl. Acad. Press, Washington, DC.

Winker, K., D.W. Warner, and A.R. Weisbrod. 1991. Unprecedented stopover site fidelity in a Tennessee Warbler. Wilson Bull. 103:512-514.

Winker, K., D.W. Warner, and A.R. Weisbrod. 1992. The Northern Waterthrush and Swainson's Thrush as transients at a temperate inland stopover site. In: J.M. Hagan, III and D.W. Johnston (eds.), Ecology and Conservation of Neotropical Migrant Landbirds. Smithsonian Inst. Press, Washington, DC, pp. 384-402.

Winkler, D.W. 1992. Causes and consequences of variation in parental defense behavior by Tree Swallows. Condor 94:502-520.

Winternitz, B.L. 1976. Temporal change and habitat preference of some montane breeding birds. Condor 78:383-393.

Wishart, R.A., and S.G. Sealy. 1980. Late summer time budget and feeding behavior of Marbled Godwits *(Limosa fedoa)* in southern Manitoba. Can. J. Zool. 58:1277-1282.

Withers, P.C. 1977. Energetic aspects of reproduction by the Cliff Swallow. Auk 94:718-725.

Woffinden, N.D., and J.R. Murphy. 1989. Decline of a Ferruginous Hawk population: a 20-year summary. J. Wildl. Manage. 53:1127-1132.

Wolf, L., E.D. Ketterson, and V. Nolan, Jr. 1991. Female condition and delayed benefits to males that provide parental care: a removal study. Auk 108:371-380.

Wolf, L., R.M. Lejnieks, and C.R. Brown. 1985. Temperature fluctuations and nesting behavior of Rock Wrens in a high-altitude environment. Wilson Bull. 97:385-387.

World Resources Institute. 1989. World Resources 1988–89. Basic Books, NY.

Wunder, B.A. 1979. Evaporative water loss from birds: effects of artificial radiation. Comp. Biochem. Physiol. 63A:493-494.

Yahner, R.H. 1983. Seasonal dynamics, habitat relationships, and management of avifauna in farmstead shelterbelts. J. Wildl. Manage. 47:85-104.

Yahner, R.H. 1991. Avian nesting ecology in small even-aged aspen stands. J. Wildl. Manage. 55:155-159.

Yahner, R.H., and D.P. Scott. 1988. Effects of forest fragmentation on depredation of artificial nests. J. Wildl. Manage. 52:158-161.

Young, B.E. 1991. Annual molts and interruption of the fall migration for molting in Lazuli Buntings. Condor 93:236-250.

Zach, R., and J.B. Falls. 1975. Response of the Ovenbird (Aves: Parulidae) to an outbreak of the spruce budworm. Can. J. Zool. 53:1669-1672.

Zaias, J., and R. Breitwisch. 1989. Intra-pair cooperation, fledgling care, and renesting by Northern Mockingbirds, *Mimus polyglottos*. Ethology 80:94-110.

Zammuto, R.M., E.C. Franks, and C.R. Preston. 1981. Factors associated with the interval between feeding visits in brood-rearing Chimney Swifts. J. Field Ornithol. 52:134-139.

Zeleny, L. 1976. The Bluebird—How You Can Help Its Fight for Survival. Indiana Univ. Press, Bloomington.

Zimmerman, J.L. 1982. Nesting success of Dickcissels *(Spiza americana)* in preferred and less preferred habitats. Auk 99:292-298.

Zink, R.M. 1986. Patterns and Evolutionary Significance of Geographic Variation in the *schistacea* Group of the Fox Sparrow *(Passerella iliaca)*. Ornithol. Monogr. No. 40.

INDEX
(Page numbers in **boldface** refer to species accounts.)

Abies lasiocarpa, 12
Accipiter cooperii, **44**
    gentilis, **45**
    striatus, **43**
Acer negundo, 20
Actitis macularia, **58**
Aeronautes saxatalis, **74**
Agelaius phoeniceus, **175**
Ammodramus bairdii, **164**
    savannarum, **165**
Amphispiza belli, **161**
Anthus rubescens, **124**
    spragueii, **125**
Aquila chrysaetos, **50**
Archilochus alexandri, **76**
    colubris, **75**
Artemisia sp., 26
ash, green, 20
Asio flammeus, **68**
    otus, **67**
aspen, 16

Bartramia longicauda, **59**
Blackbird, Brewer's, 18, 31, 36, 38, **178**, 191
    Red-winged, 31, 36, 38, **175**, 191
    Yellow-headed, 36, 38, **177**, 191
Bluebird, Eastern, 35, 38, **113**, 188
    Mountain, 14, 20, 35, 38, **115**, 188
    Western, 20, 35, 38, **114**, 188
Bobolink, 25, 36, 37, **174**, 192
Bombycilla cedrorum, **126**
boxelder, 20
Bromus tectorum, 28
brood parasitism, 1, 2, 5, 9, 10, 23
Bunting, Lark, 32, 36, 38, **163**, 191
    Indigo, 36, 38, **150**, 190
    Lazuli, 18, 20, 31, 36, 38, **149**, 190
Bureau of Land Management, 7
Buteo jamaicensis, **48**
    platypterus, **46**
    regalis, **49**
    swainsoni, **47**

Calamospiza melanocorys, **163**
Calcarius mccownii, **172**
    ornatus, **173**
    pinus, **182**
    tristis, **183**
Carpodacus cassinii, **181**
cat, domestic, 1
catbird, Gray, 36, 38 **121**, 188
Cathartes aura, **40**
Catharus fuscescens, **117**
    guttatus, **119**
Catoptrophorus semipalmatus, **57**
Certhia americana, **107**
Ceryle alcyon, **80**
Chaetura pelagica, **71**
    vauxi, **72**
Charadrius montanus, **55**
    vociferus, **56**
Chat, Yellow-breasted, 18, 20, 31, 36, 38, **146**, 190
cheatgrass, 28
Chickadee, Mountain, 11
    Black-capped, 19, 21
Chondestes grammacus, **160**
Chordeiles minor, **69**
Circus cyaneus, **42**
Cistothorus palustris, **111**
    platensis, **110**
climate, 9, 16
Coccyzus americanus, **63**
    erythropthalmus, **64**
conservation, 6, 9, 16
Conservation Reserve Program (CRP), 26
Contopus borealis, **85**
    sordidulus, **86**
    virens, **87**
Cowbird, Brown-headed, 1, 2, 9, 11, 12, 18, 19, 27, 30, 31, 32
Creeper, Brown, 10, 11, 12, 13, 20, 30, 35, 38, **107**, 187
Crow, American, 1
Cypseloides niger, **73**
Cuckoo, Black-billed, 35, 37, **64**, 185
    Yellow-billed, 31, 35, 37, **63**, 185
Curlew, Long-billed, 27, 32, 35, 38, **60**, 185

Dendroica coronata, **137**
    petechia, **136**
    townsendi, **138**
Dickcissel, 36, 37, **151**, 190
Dolichonyx oryzivorus, **174**
Dove, Mourning, 11, 12, 22, 30, 31, 32, 35, 37, **62**, 185
Dumetella carolinensis, **121**

Eagle, Golden, 27, 35, 38, **50**, 184
Elaeagnus angustifolia, 18
    commutata, 24
elm, American, 20
Empidonax hammondii, **89**
    minimus, **88**
    oberholseri, **90**
    occidentalis, **92**
    traillii, **91**
Eremophila alpestris, **99**
Euphagus cyanocephalus, **178**

Falco columbarius, **52**
    mexicanus, **54**
    peregrinus, **53**
    sparverius, **51**
Falcon, Peregrine, 35, 37, **53**, 184
    Prairie, 35, 38, **54**, 185
Finch, Cassin's, 11, 12, 30, 36, 38, **181**, 192
fir:
    Douglas, 10-12

fire:
    subalpine, 12
    avian response, 14
    in grasslands, 23
    in shrubsteppe, 28
    prescribed, 9, 14
    post-fire, 14
    suppression, 4, 5, 9, 24, 26
Flicker, Northern, 11, 30
flood control, 5, 17
Flycatcher, Cordilleran, 20, 35, 37, **92**, 187
    Dusky, 11, 12, 19, 30, 35, 38, **90**, 186
    Empidonax, 10, 12
    Hammond's, 11, 12, 30, 35, 37, **89**, 186
    Great Crested, 35, 37, **95**, 187
    Least, 35, 37, **88**, 186
    Olive-sided, 14, 19, 35, 37, **85**, 186
    Willow, 18, 31, 35, 37, **91**, 187
foraging, 9, 16
forest:
    coniferous, 5, 9-13, 29, 30
    corridor, 1, 2, 17
    deciduous, 5, 16-17
    deciduous non-riparian draws, 20-21
    deforestation, 3
    eastern deciduous, 1, 4, 5, 10
    edge, 1, 2
    fragmentation, 1, 2, 3, 5, 9, 13, 19
    interior, 1, 2
    landscape, 5, 10
    non-coniferous, 12-13
    non-riparian aspen woodland, 18-20
    old-growth, 5, 9, 10, 11, 12, 29, 30
    riparian woodland, 5, 17-18, 31
    second growth, 5, 6
    shelterbelts, 21
    stands, 10, 11, 19
Fraxinus pennsylvanica, 20

Geothlypis trichas, **144**
Godwit, marbled, 25, 35, 37, **61**, 185
Goldfinch, American, 21, 36, 38, **183**, 192
Goshawk, Northern, 11, 12, 30, 35, 38, **45**, 184
Grackle, Common, 1, 22
grassland, 5, 22-24, 184-192
grazing:
    effects, 14-15, 24-25
    livestock, 5, 14, 15, 17, 18, 20, 21, 23, 26, 27
    riparian woodland, 31
Grosbeak, Black-headed, 36, 37, **153**, 190
    Rose-breasted, 36, 37, **152**, 190

habitats:
    occurrence by species, 184-192
Harrier, Northern, 24, 27, 35, 37, **42**, 184
harvest:
    agriculture/cropland, 6
    clearcutting, 6, 9, 11, 13, 15, 19, 30
    rotation age, 30
    selection/shelterwood, 9, 11
    thinning, 9, 30
    timber, 5, 9, 17
Hawk, Broad-winged, 35, 37, **46**, 184
    Cooper's, 10, 20, 35, 37, **44**, 184
    Ferruginous, 27, 35, 38, **49**, 184
    Sharp-shinned, 10, 35, 37, **43**, 184
    Swainson's, 35, 37, **47**, 184
hemlock, western, 13
herbicides, 15, 27
High Desert Ecological Research Institute, 216
Hirundo pyrrhonota, **105**
    rustica, **106**
hole-nesting, 12, 20
Hummingbird, Black-chinned, 35, 37, **76**, 186
    Broad-tailed, 35, 37, **78**, 186
    Calliope, 12, 31, 35, 37, **77**, 186
    Ruby-throated, 35, 37, **75**, 186
    Rufous, 30, 35, 37, **79**, 186

Icteria virens, **146**
Icterus galbula, **180**
    spurius, **179**
irrigation, 5, 17

Jay, Blue, 1
Junco hyemalis, **171**
Junco, Dark-eyed, 12, 13, 14, 20, 21, 30, 36, 38, **171**, 191

Kestrel, American, 35, 37, **51**, 184
Killdeer, 18, 31, 32, 35, 37, **56**, 185
Kingbird, Cassin's, 35, 37, **96**, 187
    Eastern, 35, 37, **98**, 187
    Western, 35, 37, **97**, 187
Kingfisher, Belted, 35, 37, **80**, 186
Kinglet, Ruby-crowned, 11, 12, 13, 20, 30, 35, 37, **112**, 188
    Golden-crowned, 12, 19, 30, 112

Lanius ludovicianus, **127**
larch, western, 13
Larix occidentalis, 13
Lark, Horned, 25, 27, 32, 35, 38, **99**, 187
Limosa fedoa, **61**
Longspur, Chestnut-collared, 25, 32, 36, 38, **173**, 191
    McCown's, 25, 32, 36, 38, **172**, 191

management, 9, 16, 33
Martin, Purple, 35, 37, **100**, 187
Meadowlark, Western, 11, 21, 23, 27, 32, 36, 38, **176**, 191
Melanerpes lewis, **81**
Melospiza georgiana, **169**
    lincolnii, **168**
    melodia, **167**
Merlin, 35, 37, **52**, 184
Mimus polyglottos, **122**
Mniotilta varia, **139**
Mockingbird, Northern, 36, 38, **122**, 188
Myadestes townsendi, **116**
Myiarchus crinitus, **95**

nest:
    predation, 1, 2, 5, 9, 10, 23
    requirements, 9, 16
    success, 1, 24
Nighthawk, Common, 35, 37, **69**, 185
North American Breeding Bird Survey (BBS), 2, 8, 33

Numenius americanus, 60
Nuthatch, Red-breasted, 11, 12, 13, 19, 30

olive, 18
Oporornis tolmiei, 143
opossum, 1
Oreoscoptes montanus, 123
Oriole, Northern, 36, 37, **180**, 191
    Orchard, 36, 37, **179**, 191
Osprey, 35, 37, **41**, 184
Otus flammeolus, 65
ovenbird, 36, 38, **141**, 189
Owl, Burrowing, 27, 35, 37, **66**, 185
    Flammulated, 12, 35, 37, **65**, 185
    Long-eared, 20, 35, 38, **67**, 185
    Short-eared, 25, 27, 31, 35, 38, **68**, 185

Pandion haliaetus, **41**
Passerculus sandwichensis, **162**
Passerella iliaca, **166**
Passerina amoena, **149**
    cyanea, **150**
pest management, 15, 25
Phalaenoptilus nuttallii, **70**
Pheucticus ludovicianus, **152**
    melanocephalus, **153**
Phoebe, Eastern, 35, 38, **93**, 187
    Say's, 35, 38, **94**, 187
pine:
    lodgepole, 14
    ponderosa, 10-12
Picea engelmannii, 12
Pinus ponderosa, 12
    contorta, 14
Pipilo chlorurus, **154**
    erythrophthalmus, **155**
Pipit, American, 36, 38, **124**, 188
    Sprague's, 25, 32, 36, 38, **125**, 188
Piranga ludoviciana, **148**
    olivacea, **147**
Plover, Mountain, 25, 35, 38, **55**, 184
Pooecetes gramineus, **159**
Poorwill, Common, 35, 38, **70**, 185
population, 1, 2, 3, 4, 5, 6, 9, 16, 17, 27
Populus tremuloides, 16
power generation, 17, 20
Progne subis, **100**
Pseudotsuga menziesii, 10-12

raccoon, 1
recreation, 17
Redstart, American, 20, 36, 37, **140**, 189
Regulus calendula, **112**
Riparia riparia, **104**
road development, 3
Robin, American, 11, 13, 14, 18, 19, 21, 22, 31, 36, 38, **120**, 188

sagebrush, 26
Salix sp., 16
Salpinctes obsoletus, **108**
Sandpiper, Upland, 25, 32, 35, 37, **59**, 185
    Spotted, 35, 37, **58**, 185

Sapsucker, Red-naped, 12, 19, 20, 31, 35, 37, **83**, 186
    Williamson's, 10, 12, 20, 30, 35, 38, **84**, 186
    Yellow-bellied, 35, 37, **82**, 186
Sayornis phoebe, **93**
    saya, **94**
Seiurus aurocapillus, **141**
    motacilla, **142**
Selasphorus platycercus, **78**
    rufus, **79**
Setophaga ruticilla, **140**
Shrike, Loggerhead, 36, 38, **127**, 189
shrubsteppe, 5, 26-28, 32
Sialia currucoides, **115**
    mexicana, **114**
    sialis, **113**
silverberry, 24
silviculture, 9, 10, 12, 13
Siskin, Pine, 13, 20, 30, 36, 38, **182**, 192
snags, 11, 12, 15, 16, 18
snowberry, 24
Solitaire, Townsend's, 35, 38, **116**, 188
Sparrow, Baird's, 25, 32, 36, 38, **164**, 191
    Brewer's, 26, 27, 32, 36, 38, **158**, 190
    Chipping, 11, 13, 14, 30, 36, 38, **156**, 190
    Clay-colored, 23, 32, 36, 38, **157**, 190
    Field, 21
    Fox, 13, 15, 18, 36, 38, **166**, 191
    Grasshopper, 23, 25, 27, 32, 36, 38, **165**, 191
    House, 22
    Lark, 20, 25, 32, 36, 38, **160**, 190
    LeConte's, 25
    Lincoln's, 15, 18, 19, 31, 36, 38, **168**, 191
    Sage, 26, 27, 36, 38, **161**, 190
    Savannah, 18, 23, 25, 27, 32, 36, 38, **162**, 190
    Sharp-tailed, 25
    Song, 36, 38, **167**, 191
    Swamp, 36, 38, **169**, 191
    Vesper, 11, 14, 26, 27, 30, 32, 36, 38, **159**, 190
    White-crowned, 11, 18, 30, 36, 38, **170**, 191
Speotyto cunicularia, **66**
Sphyrapicus nuchalis, **83**
    varius, **82**
    thyroideus, **84**
Spiza americana, **151**
Spizella breweri, **158**
    pallida, **157**
    passerina, **156**
spruce, 12, 19
Starling, European, 31
Stelgidopteryx serripennis, **103**
Stellula calliope, **77**
Sturnella neglecta, **176**
survival, 9, 16
Swallow, Bank, 35, 37, **104**, 187
    Barn, 35, 37, **106**, 187
    Cliff, 35, 37, **105**, 187
    Northern Rough-Winged, 35, 37, **103**, 187
    Tree, 13, 19, 30, 35, 37, **101**, 187
    Violet-green, 35, 37, **102**, 187
Swift, Black, 35, 37, **73**, 185
    Chimney, 35, 37, **71**, 185
    White-throated, 35, 37, **74**, 186
    Vaux's, 12, 30, 35, 37, **72**, 185

Symphoricarpos occidentalis, 24

Tachycineta bicolor, **101**
    thalissina, **102**
Tanager, Scarlet, 36, 37, **147**, 190
    Western, 11, 14, 20, 36, 37, **148**, 190
Thrasher, Sage, 26, 27, 36, 38, **123**, 188
Thrush, Hermit, 11, 12, 13, 20, 30, 36, 37, **118**, 188
    Swainson's, 11, 12, 13, 19, 21, 30, 31, 36, 37, **118**, 188
    Varied, 12, 13
Towhee, Green-tailed, 26, 36, 38, **154**, 190
    Rufous-sided, 21, 36, 38, **155**, 190
Troglodytes aedon, **109**
Tsuga heterophylla, 13
Turdus migratorius, **120**
Tyrannus tyrannus, **98**
    verticalis, **97**
    vociferans, **96**

Ulmus americana, 20
United States Fish & Wildlife Service, 7
United States Forest Service (USFS), 7, 9
USDA Animal Plant Health Inspection Service (APHIS), 25, 26

Vermivora celata, **134**
    peregrina, **133**
    ruficapilla, **135**
Veery, 18, 31, 36, 37, **117**, 188
Vireo bellii, **128**
    flavifrons, **130**
    gilvus, **132**
    olivaceus, **131**
    solitarius, **129**
Vireo, Bell's, 36, 37, **128**, 189
    Red-eyed, 20, 36, 37, **131**, 189
    Solitary, 10, 11, 14, 20, 30, 36, 38, **129**, 189
    Yellow-throated, 36, 38, **130**, 189
    Warbling, 11, 19, 31, 36, 37, **132**, 189

Vulture, Turkey, 35, 37, **40**, 184

Warbler, Black-and-White, 20, 36, 38, **139**, 189
    MacGillivray's, 12, 18, 19, 31, 36, 37, **143**, 189
    Nashville, 15, 18, 36, 38, **135**, 189
    Orange-crowned, 19, 21, 36, 38, **134**, 189
    Tennessee, 36, 37, **133**, 189
    Townsend's, 11, 12, 13, 30, 36, 38, **137**, 189
    Wilson's, 18, 21, 31, 36, 38, **145**, 190
    Yellow, 31, 36, 38, **136**, 189
    Yellow-rumped, 11, 13, 14, 19, 30, 36, 38, **137**, 189
Waterthrush, Northern, 36, 38, **142**, 189
Waxwing, Cedar, 11, 30, 36, 38, **126**, 188
Willet, 25, 35, 37, **57**, 185
willow, 16
Wilsonia pusilla, **145**
wintering areas:
    of species, 33-38
    problems, 3-4, 6
wolfberry, 24
Woodpecker, Black-backed, 14
    Hairy, 30
    Lewis', 35, 38, **81**, 186
    Pileated, 11, 12, 30
Wood-Pewee, Eastern, 35, 37, **87**, 186
    Western, 19, 31, 35, 37, **86**, 186
Wren, House, 13, 19, 20, 30, 31, 35, 38, **109**, 188
    Marsh, 35, 38, **111**, 188
    Rock, 35, 38, **108**, 187
    Sedge, 35, 38, **110**, 188
    Winter, 11, 30

Xanthocephalus xanthocephalus, **177**

Yellowthroat, Common, 18, 31, 36, 38, **144**, 189

Zenaida macroura, **62**
Zonotrichia leucophrys, **170**

THE HIGH DESERT ECOLOGICAL RESEARCH INSTITUTE

The High Desert Ecological Research Institute, under the direction of Dr. David S. Dobkin, was established in 1993 to serve as a regional center for ecological research and policy analysis with a primary focus on natural resource issues related to the Inter-Mountain West. The Institute conducts cooperative, multidisciplinary, long-term research on a wide range of problems in natural resource management and conservation in western North America.

The High Desert Ecological Research Institute is an independent, non-profit organization.